NETWORK NEUTRALITY

Gabriele Balbi

with Simone Fari, Giuseppe Richeri and Spartaco Calvo

NETWORK NEUTRALITY

Switzerland's role in the genesis of the Telegraph Union, 1855-1875

PETER LANG

Bern · Berlin · Bruxelles · Frankfurt am Main · New York · Oxford · Wien

Bibliographic information published by die Deutsche Nationalbibliothek
Die Deutsche Nationalbibliothek lists this publication in the Deutsche Nationalbibliografie; detailed bibliographic data is available on the Internet at ‹http://dnb.d-nb.de›.

British Library Cataloguing-in-Publication Data: A catalogue record for this book is available from The British Library, Great Britain

Library of Congress Cataloging-in-Publication Data is available.

This book is a collective effort of the research group involved in the Swiss National Fund project "The Swiss Influence in the ITU´s decision-making process 1865-1914". Hypotheses, sources and ideas were discussed collectively, while the chapters were written individually. Giuseppe Richeri wrote the Introduction, Spartaco Calvo Chapter 1, Simone Fari Chapter 6, and Gabriele Balbi Chapter 2, 3, 4, 5 and the Conclusion.

Earlier versions of some segments have already appeared in academic journals. Specifically, a previous version of chapter 1 was published in French in the Revue Suisse d'Histoire 61(4) in 2011; some elements of chapter 2 appeared in ICON 15 in 2009 and an abstract of the main topics of this book was published both in Italian and in English in *Tst: Transportes, Servicios y telecomunicaciones 25 and Journal of European Integration History* 19(2) in 2013.

Translated by Patricia Kennan.

Printed with the support of the Fondazione Hilda e Felice Vitali, Lugano (Switzerland).

Cover: Monument celebrating the Telegraph Union's foundation, inaugurated in Berne in 1922.

ISBN 978-3-0343-1527-2 pb.   ISBN 978-3-0351-0744-9 eBook

Hochfeldstrasse 32, CH-3012 Bern, Switzerland
info@peterlang.com, www.peterlang.com

Printed in Switzerland

Contents

Acknowledgements

Over the years in which we have been involved working on such a specific, complicated and complex research we have collected many debts of both an intellectual and material level. We would like to thank in particular Verdiana Grossi, Andrea Giuntini, Jakob Tanner, David Gugerli, all the scholars on the panel at the SHOT Conference in Cleveland 2011 (Richard R. John, Peter A. Shulman, Heidi J. S. Tworek, Andrew J. Butrica and Graeme J. N. Gooday) and in a workshop held at Maastricht in 2012 (especially Andreas Fickers and Pascal Griset) for their precious advice. Without the help and readiness of kind archivists this book would perhaps be different – and we feel worse. We have debts with Ronny Trachsel, Madeleine Burri and Heike Bazak of the Historic Archives and Library of the PTT, Juri Jaquemet of the Museum für Kommunikation in Berne, Christine Lauener of the Swiss Federal Archives, Heather Heywood of the ITU Archives and all the personnel of the Biblioteca Universitaria di Lugano. Particular thanks to Rita Deiana Brügger, for her attentive bio-bibliographical research on some of the main figures in the research. Always in Lugano, we would like to thank the Research Service USI-SUPSI, in the phase preceding thc writing of the project and for their help in submitting it. Our greatest debt is to the Swiss National Fund, which financed our project right from the beginning. Without its help and backing this research would certainly never have seen the light.

Introduction

The long-term project

In 2009 a research group at the Institute of Media and Journalism of the Università della Svizzera italiana, Lugano, set up a long-term project entitled "The Swiss Influence in the ITU's decision-making process 1865–1914". As the title suggests, the aim was to analyze the role played by Switzerland in the creation, early years of development and decision-making processes of the Telegraph Union (nowadays known as the International Telecommunication Union), bringing together political, economic, technical and cultural points of view. The first years of research carried out by the closely-knit team composed of Gabriele Balbi, Simone Fari, Giuseppe Richeri and Spartaco Calvo resulted in the presentation of numerous papers at conferences and publication in scientific journals[1]. The rationale of the present volume is partly to collect all these experiences into an organic whole while at the same time drawing on the debates and discussions coming out of the various conferences, suggestions for changes from various peer reviews, as well as the support of a considerable amount of inedited material.

1 G. Balbi et al., "Swiss specialties: Switzerland's role in the genesis of the Telegraph Union, 1855–1875," *Journal of European Integration History*, 19/2 (2013); G. Balbi et al., "Specialità svizzere. L'influenza della Confederazione elvetica sull'origini dell'Unione Telegrafica, 1855–1875," *Tst: Transportes, Servicios y telecomunicaciones*, 25(2013); S. Calvo et al., "La voie suisse aux télécommunications. Politique, économie, technologie et société (1850–1915)," *Revue Suisse d'Histoire* 61/4 (2011); G. Balbi et al., "'Bringing together the two large electric currents that divide Europe': Switzerland's Role in Promoting the Creation of a Common European Telegraph Space, 1849–1865," *ICON* 15 (2009).

Methodology and main sources

The main methodology backing our approach is multifocal, i.e. it holds that in order to understand fully the social construction of the media and telecommunications, equal consideration must be given to the political, economic, technical and social features contributing to the formation of the object in study. A methodology like this is particularly appropriate when analysing telecommunications, complex structures calling for careful and multiform reflection by all the social components that regulate, realize and use them[2]. It needs to be said, too, that the reference discipline governing this volume is the political economy of communication, which analyses how the structuring presence of the state influences business strategies or, in other words, how constitutive choices made by politics influence media development (Starr 2004, John 2010). The political economy of communication tends to be associated with national politics because states often have different ideas about the ways in which they want to control, develop or limit communications.

This project aims to interface a political economy background with international politics and identify European constitutive choices concerning telecommunications. We hypothesize that Switzerland carried out a decisive role in stimulating and guiding established objectives, i.e. the creation of an international organization able to regulate telecommunications traffic – or in our period essentially electric telegraphy – on a continental level.

The historical analysis was carried out on both primary and secondary sources. A fundamental role was indeed played by the correspondence and commissions set up by the Swiss Federal Council,

2 N. Rosenberg, *Exploring the black box: technology, economics, and history* (Cambridge UK and New York: Cambridge University Press, 1994); C. Sterling, P. Bernt and M. B. H. Weiss, *Shaping American telecommunications: a history of technology, policy and economics* (Mahwah, New York: Lawrence Erlbaum Associates, 2006); G. Balbi, "Studying the Social History of Telecommunications. Between Anglophone and Continental Tradition," *Media History*, 15/1, 2009.

whose federal papers we consulted with care[3]. Also important were the Swiss government and parliament documents on the international conferences and reunions on telegraphy which took place in various seats between the mid-1850s and mid-1870s, now available in the Swiss Federal Archives, Bern[4]; documents of the Swiss Post and Telegraph Department conserved in the PTT History Archive and Library in Bern[5]; lastly the minutes of the international telegraph conferences and correspondence among the representatives of the Telegraph Union, conserved at the ITU Library and Archives in Geneva[6].

Next to the primary sources guiding our reconstruction of the story of the telegraph, we quite naturally turned to secondary sources. However, with the exception of two articles by Verdiana Grossi[7], Switzerland's role in thc process of creation and institutionalization of the Union has been almost ignored and therefore there little help was forthcoming. However, there are works which have investigated the role that other nations played in the international organization of telecommunications, most particularly France and the United States[8]. This book is therefore a

---

3 See <http://www.amtsdruckschriften.bar.admin.ch/setLanguage.do?lang=FR&currWebPage=searchHome>, last access 3 November 2013.

4 See <http://www.bar.admin.ch/index.html?lang=en>, last access 3 November 2013.

5 See <http://www.mfk.ch/pttarchivbibliothek.html?&L=2>, last access 3 November 2013.

6 See <http://www.itu.int/en/history/Pages/LibraryAndArchives.aspx>. For the typologies of documents kept in this archive see a note by Gabriele Balbi and Simone Fari available on <http://unhistoryproject.org/research/research_experiences-balbi.html>, last access 3 November 2013.

7 See V. Grossi, "Le rôle international de personnalités suisses du XIXe siècle dans le domaine des télégraphes," *Hispo,* Octobre 1984 and "Technologie et diplomatie suisse au XIXe Siècle," *Relations internationales* 39 (1984).

8 In the case of France see L. Laborie (*La France, l'Europe et l'ordre International des communications, 1865–1959*, Thèse pour le Doctorat en histoire contemporaine, Université Paris IV-Sorbonne, 2006), while the works of M. L. B. Feldman (*The United States in the International Telecommunication Union and in Pre-ITU conferences: Submarine Cables, Overland Telegraph, Sea and Land Radio, Telecommunications* (Baton Rouge: printed autonomously, 1974) and A. Rutkowski & W. P. Dizard ("The International Telecommunication Union and the United States: partners of rivals?" In *International Telecommunications and information Policy.*

first attempt to collocate Switzerland – then a new starter – at the centre of an analysis of the origins of the first international organization.

Reference literature

Besides filling a scientific gap in evaluating Switzerland's influence on the Telegraph Union, this volume also intends to contribute to a wider-reaching scientific literature dealing principally with the history of European communications and institutions regulating them.

Firstly, our research marks a further step towards the creation of a European media history and seeks to integrate in some ways with recent researches on European television history[9]. Though European media history is an emerging field of studies[10], years ago the crucial role of communication infrastructures in "Networking Europe" was placed at the heart of the matter by a group of technology historians, gravitating around the "Tensions of Europe Association"[11]. In particular, this work

Washington: Communications Pr., 1984) witness the American contribution, especially after WWII.

9 See for example J. Bignell and A. Fickers (eds.), *A European television history* (Malden MA and Oxford: Wiley-Blackwell, 2008); J. Bourdon, *Du service public à la télé-réalité. Une histoire culturelle des télévisions européennes, 1950–2010* (Paris: INA Éd. 2011).

10 On the increasing attention on European media history see S. Kinnebrock, C. Schwarzenegger and E. McLuskie (eds.), *What is Communication History? European Answers II*. Special Issue of *Medien & zeit*, 3–4 (2011). In 2009 ECREA, the most important association for the study of European communication set up a Communication History Section dedicated to the diachronic study of communication in Europe.

11 On this group, see <http://www.histech.nl/tensions>. One of the most interesting books on the topic is: E. Van der Vleuten and A. Kaiser (eds.), *Networking Europe. Transnational Infrastructures and the Shaping of Europe, 1850–2000* (Sagamore Beach, MA: Science History Publications, 2006). Forthcoming: A. Fickers and P. Griset. *Eventing Europe: Electronic Information and Communication Spaces in Europe, 1850–2000* (London: Palgrave/MacMillan, 2014).

group sees telecommunications not only as an infrastructure network but also as a decisive tool in developing the so-called "hidden integration" process[12] which some hold to have begun in Europe back in the nineteenth century, way back therefore before the idea of any European Union came to the fore. Among the material structures influencing the practices and definitions of Europe, telecommunications and particularly the telegraph hold indeed an important position, while the Telegraphic Union is recognized as the first institution to conceive the idea of a European space. Investigating the origins of this institution means therefore tracing the idea of European integration via communication, which as this volume shows, is contained in a Swiss project of the early 1850s.

Other authors in "Networking Europe" see a fundamental role of smaller nations as linking agents in Europe, not only in the realm of communications, especially because of the recognized experience of their technical cadres[13]. Our book in a certain sense offers a backing to this viewpoint, reflecting over the importance and excellence of the Swiss technocrats, especially Louis Curchod, who became the first head of the International Bureau of the Telegraph Administration, the managing body of the International Telegraph Union.

Linked again with the "Tensions of Europe" tradition, there is a second field of research which this book aims to be part of – the transnational approach. The term "transnational" has a long, complex

12 T. J. Misa and J. Schot, "Inventing Europe: Technology and the Hidden Integration of Europe. Introduction to the special issue," *History and Technology*, 21/1 (2005).

13 "The more specialized arenas of technical experts, such as the annual meetings of international organizations, can also be read as European events, where power is expressed through performance. Representatives of smaller or less powerful nations in particular have often been able to use the performances of technical expertise or European unity to position themselves more centrally on the European stage" (A. Badenoch and A. Fickers, "Introduction: Europe materializing? Toward a transnational history of European infrastructures," in *Materializing Europe: transnational infrastructures and the project of Europe*, ed. A. Badenoch and A. Fickers (Basingstoke, Hampshire and New York: Palgrave Macmillan, 2010), 15.

and contested history in many political and academic contexts[14], not excluding the history of technology[15]. A transnational approach to the study of infrastructures such as communication networks deals with flows over national borders, the international institutions regulating them, supranational economic powers, the circulation of ideas and people through media that cannot be circumscribed by national spaces. Our text aims at reconstructing an unprecedented history of the first transnational institution dealing with regulating communication between various nations and also showing the masterly way the Swiss regulated and ran these flows of transnational communication.

A third developing strand of research into which our text can be drawn is that of the history of international organizations[16], in particular in their functions as political/economic coordinators and technological/social standardisers. Here our focus is on non-governmental institutional players capable of guiding political bodies like national governments or economic ones like big companies. Though the Telegraph Union was the first ever international organization, it has had scarce mention in this field. There are indeed the works of a commemorative nature published by the Union itself, but historical objectivity often takes a back seat to the self-celebratory reconstruction of events[17]. The most important

14 A. P. Iriye and Y. Saunier (eds.), *The Palgrave dictionary of transnational history* (Basingstoke UK and New York: Palgrave Macmillan, 2009); C. Conrad, "Social policy history after the transnational turn," in *Beyond Welfare State Models. Transnational Historical Perspectives on Social Policy*, ed. P. Kettunen and K. Petersen (Cheltenham, UK and Northampton, MA: Edward Elgar, 2011).

15 E. Van der Vleuten, "Towards a transnational history of technology: meanings, promises, pitfalls," *Technology and Culture,* 49/4 (2008).

16 M. Herren, *Internationale Organisationen seit 1865. Eine Globalgeschichte der internationalen Ordnung* (Darmstadt: Wissenschaftliche Buchgesellschaft, 2009); S. Kott (dir.), *Une autre approche de la globalisation: Socio-histoire des organisations internationales (1900–1940*). Numéro spécial de *Critique internationale*, 52 (2011).

17 ITU, *L'Union télégraphique international: 1865–1915* (Berne: Bureau international de l'Union télégraphique, 1915); ITU, *L'Union Internationale des télécommunications, cent ans de coopération internationale* (Geneve, 1965); A. R. Michaelis, *Du sémaphore au satellite* (Geneve: ITU, 1965); *L'Union internationale des télécommunications*, 1865–1995 (London: International Systems

scientific studies on the Telegraph Union are, instead, by law and political sciences scholars focusing above all on the internal functioning of the Union itself[18]. Only in more recent times has the attention of scholars turned to the institutions of the Union, its functions and more importantly for our discourse the international regulation of telecommunications[19]. Achieving technical, regulatory and tariff standardization, was indeed a major accomplishment, with economic, social and technical effects which so far scholars have all but ignored. Our book hopes to redress at least in part this situation by studying Switzerland's role in creating a transnational institution capable of regulating communications on a European and then world level.

---

and Communications LTD, 1995); P. Tarjanne (avant-propos par), *Union internationale des télécommunications: célèbre 130 années d'existence 1865–1995* (Genève: International Telecommunication Union and International Systems and Communications Ltd., 1996).

18 V. Meyer, *L'Union Internationale des Telecommunications et son Bureau* (Bern: manuscript, 1946); A. G. Codding Jr, *The International Telecommunication Union. An Experiment in international cooperation* (Leiden: E. J. Brill, 1952); J. F. Navarro, *ITU: síntesis de organización, funcionamiento y objetivos de la Union Internacional de Telecomunicaciones* (Chile: Entel-Chile, 1975); J. Horrenberger, *L'Union Internationale des Télécommunications ou Les exigences techniques comme factor de la coopération internationale* (Mémoire pour l'obtention du diplôme des Hautes Etudes Européens section des sciences de l'information, Université de Strasbourg, 1976); P. Durand Barthez, *Union Internationale des Télécommunications* (Thèse pour le doctorat en droit, Université de Paris I-Pantheon-Sorbonne Sciences Economiques-Science Humaines-Sciences Juridiques, 1979); F. Lyall, *International communications: the International Telecommunication Union and the Universal Postal Union* (Farnham, Surrey UK and Burlington, VT: Ashgate, 2011).

19 G. Nachszunow, *Développement des télécommunications et organisations internationales* (Créteil: W. Nachszunow, 1989); S. Schmidt and R. Werle, *Coordinating Technology: Studies in the International Standardization of Telecommunications* (Cambridge-London: MIT Press, 1998); Laborie, *La France, l'Europe et l'ordre international des communications (1865–1959)*; L. Laborie, *L'Europe mise en réseaux. La France et la coopération internationale dans les postes et les télécommunications, années 1850-années 1950* (Bruxelles: PIE Peter Lang, 2010); S. Fari, *Una penisola in comunicazione. Il servizio telegrafico italiano dall'Unità alla Grande Guerra* (Bari: Cacucci Editore, 2008).

The contents of the book

The book contains six chapters. Chapter 1 analyses the Swiss approach to telecommunications during the nineteenth century, highlighting features of political economy, economics and the technological background as well as the Swiss promotion of the system of telecommunications on a national level. Chapter 2 examines the ten years leading up to the creation of the Telegraph Union, the early need to communicate on an international level and the formation of the two unions, which thanks to Swiss mediation converged and merged into the Telegraph Union. It was right in the mid-fifties that the Federal Council and some top telegraph managers thought up the idea of a common European space. Chapter 3 tells of the foundation of the TU at the 1865 Paris International Conference and the leading role Switzerland obtained within the new body thanks to the skills of Louis Curchod, its delegate. Chapter 4 focuses on the 1868 Vienna International Conference which approved the Curchod/Swiss government project for the creation of an international bureau to regulate the Telegraph Union. Since the Bureau was set up at Bern, on Swiss soil, and placed under the Swiss Postal Department, it was tantamount to being under the Swiss government. And the first head was, as can be imagined, Curchod. Chapter 5 reviews the debates at the 1872 Rome Conference, particularly the attempt to snatch control of the Bureau which not only failed, but strengthened the Federal Council's hold. Chapter 6 focuses principally on the 1875 St Petersburg Conference and illustrates to perfection how the Bureau was able to influence international communications and decide the dynamics of its regulations. The conclusion rounds off the discourse with Swiss specialisms and other features of a political, economic and technical nature which all contributed to taking the country to the head of such an important institutional body.

Chapter 1
Switzerland takes on telecommunications. The politics, economics, technology and society of the period

Introduction

Any appreciation of the role Switzerland played in the creation and structure of the international telecommunication systems needs to take into full account the Swiss political economy of communication. It is crucial therefore to understand what political, economic and technical reasons made telecommunications into such a central issue in the public debate of the times and what the Federal Council's political agenda was from the 1850s right through to the new century. Investigation needs to be made into the "constitutive choices" or founding ideas of Swiss politics that affected the development of telecommunications[1], as well as the technical developments and social significances of point-to-point communications[2] at their origins.

1 On the relevance of the "constitutive choices", see P. Starr, *The creation of the media: political origins of modern communications* (New York: Basic Books, 2004).

2 In this book the expression "point-to-point communications" is used as synonymous of "telecommunications" and refers to distant communication technologies in which the transmission and reception of the message take place without the physical transmission of the same, see J. B. Thompson, *The media and modernity: a social theory of the media* (Stanford, CA: Stanford University Press, 1995). Typical examples are electric/wireless telegraphs and landline/cell phones. For a short history of telecommunications see G. Balbi, "Telecommunications," in *Handbook of Communication History*, ed. P. Simonson, J. Peck, R. T. Craig, and J. P. Jackson (London and New York: Routledge, 2012).

The aim of this chapter is therefore to analyse the particular Swiss approach to telecommunications. While the first section offers a brief analysis of the origins of telegraphy and telephony in Switzerland in the nineteenth century, the second identifies the intersections between the development of telecommunications and Swiss institutions, as well as the importance of the country's geographical position, its radically-inspired economic doctrine, the key role of foreign trade and the political clout of the technical elites. Finally, the third part deals with a series of hypotheses on the reasons which brought Switzerland to promote a European telegraph network and guide its development by means of the International Bureau of Telegraph Administrations (hereafter the Bureau).

1.1 The beginnings of telecommunications in Switzerland

The invention of the first electric instruments for long distance communication goes back to the 1820s[3] and by the late 1830s to early 1840s the most industrialised countries of the time were beginning to introduce experimental lines. Between 1837 and 1840, for example, William Cooke and Charles Wheatstone convinced some British railway companies to take up their patents, so as to facilitate service communications between stations and allow trains to run more smoothly[4].

It was, however, in the mid-forties that in both Great Britain and the States the first private companies were established to provide a public telegraph service[5]. Then within a few years (1845–1850), almost all European countries introduced the facility, though mostly as a public monopoly. Initially many countries like the Kingdom of Piedmont and

3 J. J. Fahie, *A History of Electric Telegraphy to the year 1837* (London: E & F. N. Spom, 1884).

4 J. Kieve, *The Electric Telegraph. A Social and Economic History* (Devon: David and Charles Newton Abbot, 1973).

5 K. Beauchamp, *History of Telegraphy* (IEE: Stevenage, 2001), 51–99.

Sardinia (henceforth Piedmont), France and Austria operated the service for state use alone[6]. In a later stage, given the rapid technological evolution and the commercial/industrial classes' need for rapid communications, telegraph lines spread for public use, too. Technologically they were quite simple, made of three basic elements – wooden poles, glass or ceramic insulators and iron wires. There were consequently very few differences in lines from country to country, so that right from the beginning they were automatically standardized. Telegraph apparatus was instead technologically quite different, especially in the early years[7]. In Great Britain the most important private company, the *Electric*, adopted Cooke and Wheatstone's system, which was mainly based on a magnetic needle moving when crossed by an electric current, so creating a combination of signals which together formed a communication code. However, in order not to infringe patent rights, the other British companies adopted very different telegraphs and codes. In the States Samuel Morse's electromagnetic telegraph was most commonly adopted, as in France Breguet's dial system (which originally reproduced the movements and signals of the optic telegraph)[8]. On the Italian peninsula states varied in their adoption of telegraphic apparatus: Henley was chosen in the Kingdom of Naples, Wheatstone and Morse in Piedmont and Breguet in Tuscany[9]. During the 1850s the service everywhere went through an extraordinary period of expansion, which inevitably led to the rise of the first international connections. This is the main reason why in the later part of the decade standardization increased and the Morse telegraph and code were adopted in most national and international communications.

---

6 A. A. Huurdeman, *The Worldwide History of Telecommunications* (London: IEE, 2003).

7 R. Burns, *Communications: an international history of the formative years* (London: IEE, 2004).

8 C. Bertho-Lavenir (dir.), *Histoire des télécommunications en France* (Toulouse: Eres, 1984); F. Barbier et C. Bertho-Lavenir, *Histoire des medias: de Diderot à Internet* (Paris: A. Colin, 2003).

9 S. Fari, "Technology on the wire. Technological changes in the first thirty years of the Italian telegraph experience: achievements and difficulties," in *Communication and its lines. Telegraphy in the 19th Century among economy, politics and technology*, ed. A. Giuntini (ed.) (Prato: Istituto di Studi Storici Postali, 2004).

Of all the countries undergoing industrialization in the mid-century, Switzerland was the last to install its own telegraph network, with electric telegraphy being adopted only in the early fifties. On 15 October 1851 the Department of the Post and Public Works sent the Federal Council (i.e. the Swiss Government) a report on the installation of domestic lines, proposing a call for building tenders among private companies. The government agreed, and on 1 November sent a memorandum around the cantons to sound out their interest. The project foresaw the construction of telegraph lines running north-south and east-west, following the lay-out of the already existing railway networks. The National Council Commission[10] tasked with examining the project made some changes and declared importantly that "the creation of electric telegraphs was the State's business"[11]. The Commission evidently fully recognised the political and economic utility of the proposal and therefore held that the new means of communication should be included among the publicly managed natural monopolies. On 23 December 1851, the *Loi sur la construction de télégraphes électriques* (Law on building Electric Telegraphs) containing the Commission's modifications came into effect. A year later, on 5 December 1852, Switzerland officially began the construction of a national telegraph network[12].

It needs to be held in mind that how to handle the telegraph service was being debated in all industrialized countries right from the beginning. Great Britain and the States alone chose a system of free competition and granted the service to private companies, though during the fifties and sixties there was heated debate in both countries over the possibility of an exclusive Post Office monopoly[13]. The

10 The National Council, i.e. the lower house of the Swiss Federal Assembly.

11 «la création des Télégraphes électriques est l'affaire de la Confédération». Feuille Fédérale (henceforth FF), vol. 3, n. 64 (henceforth 3/64), 27 Décembre 1851, p. 337.

12 This brief reconstruction of the origins of the electric telegraph in Switzerland is based on: Direction Générale des PTT (publ. par), *Un siècle de télécommunications en Suisse, 1852–1952. Tome I, Télégraphe* (Berne, 1952), 111–142.

13 W. S. Jevons, *On the analogy between the Post Office, Telegraphs, and other systems of Conveyance of the United Kingdom, as regards Government control*, read

Americans finally chose to leave all in private hands, thus also determining a separation between their telegraph and postal services[14]. Differently in Great Britain, because of the inefficiencies due to a mainly cartel-regulated oligopolistic market, telegraphy was nationalized and entrusted to the Post Office (1869)[15]. All the other European countries went for a public monopoly[16]. France, for example, extended the law entrusting the state with the running of optic telegraphy to the electric telegraph system, while later Germany inherited from Prussia and Italy from Piedmont a legislation making electric communications public services[17].

The telephone instead was beginning to be introduced into the States and Europe already post 1876, after Alexander Graham Bell had deposited his troubled patent and set up a company to exploit the new instrument[18].

Even in the early years of the spread of the telephone, national governments had to choose whether to grant exclusive rights to privates, keep them for the state or opt for a mixed public-private system. The first method was adopted in the States and Canada. *The Bell Telephone Company* (later *AT&T*) held a dominant position from 1880 until its patents ran out in 1893–4. Many independent companies then began to

---

on 10 April 1867 to the Manchester Statistical Society; W. S. Jevons, *Government and the Telegraphs. Statement of the case of the Electric & International Telegraph Company against The Government Bill for acquiring the Telegraphs* (London: Effingham Wilson, 1868).

14 R. John, "Private enterprise, public good? Communications deregulation as a national political issue, 1839–1851," in *Communication and its lines. Telegraphy in the 19th Century among economy, politics and technology*, ed. A. Giuntini (Prato: Istituto di Studi Storici Postali, 2004).

15 C. R. Perry, *The Victorian Post Office. The growth of a bureaucracy* (Woodbridge: The Royal Historical Society/The Boydell Press, 1992).

16 ITU Library and Archives, *Correspondances du Bureau International des Administrations Télégraphiques* (henceforth ITU-Corr), feuille n. 168/6, 21 Juin 1872.

17 *La législation télégraphique – Étude publiée par le Bureau International des Administrations Télégraphiques d'après des documents officiels* (Berne: Bureau International des Administrations Télégraphiques, 1876).

18 J. Brooks, *Telephone. The First Hundred Years* (New York & London: Harper & Row, 1972).

extend their networks by including areas that had been overlooked by Bell, so creating a lively competition that helped spread the service[19]. A state monopoly was chosen right from the beginning in Germany, Luxembourg, Greece and Romania while experimentation was carried out totally by privates, before gradually passing under state management, in France, Belgium, Austria, Hungary, Italy, Great Britain and Switzerland[20]. Other countries chose a mixed system of management financed by joint public and private capital. Urban networks were mainly granted to private companies, while national governments ran long distance lines (strategically more important). This happened, for example, in Holland and the Scandinavian countries[21].

Switzerland was quicker to experiment with the telephone than the telegraph. By mid-November 1877 the news of the establishment of a telephone line in Germany reached Switzerland and August Frey, head of the Swiss Telegraph Department, wrote at once to his German counterpart to enquire about it. Already on 17 December the Minister of Posts issued a circular letter announcing the decision to open a public telephone service and put it under the control of the Telegraph Department. As in most countries adopting a public monopoly system, the Swiss telephone service was subject to the same rules as the telegraph and was seen as following on naturally from it:

---

19 C. S. Fischer, *America Calling: A Social History of the Telephone to 1940* (Berkeley: University of California Press, 1992); R. R. John, *Network Nation. Inventing American Telecommunications* (Cambridge MA and London: The Belknap Press of Harvard University Press, 2010).

20 See P. Aulas, *Les origins du téléphone en France, 1876–1914* (Paris: ADHE, 1999), pp. 146–156; A. Hazlewood, "The Origin of the State Telephone Service in Britain," *Oxford Economic Papers, new series* 5/1 (1953); G. Balbi, *Le origini del telefono in Italia. Politica, economia, tecnologia e società* (Milan: Bruno Mondadori, 2011) or, for a brief English version, "The Origins of the Telephone in Italy, 1877–1915: Politics, Economics, Technology and Society," *International Journal of Communication*, 5 (2011).

21 C.-F. Helgesson, *Making a natural monopoly: the configuration of a techno-economic order in Swedish telecommunications* (Stockholm: Stockholm School of Economics, the Economic Research Institute, 1999); O. De Wit, *Telefonie in Nederland 1877–1940* (Den Haag: Cramwinckel, 1998), 333–334.

> The Federal Council did not hesitate for a moment to allow the inclusion under the designation "electric telegraphs" of all the installations intended to exchange by means of electricity thoughts between two points at a greater or lesser distance.[22]

Between appeals, uncertainties and second thoughts the Federal Council began to experiment on telephone lines in 1879/1880 before issuing a definitive ordinance on 29 November 1880. In July of the same year it had already granted the *Zurich Telephone Company* permission to establish a telephone network in the city. It was the first and only private telephone concern in Switzerland and was taken over by the public administration when the concession ran out in 1885[23].

Various political, economic, technological and social reasons convinced the Federal Council and institutions of the importance of telecommunications for the country's future. The remaining part of the chapter will therefore be dedicated to indepthing these motivations and introducing some theories on the reasons why Switzerland had such an interest in the communication of thought.

1.2 Federalism and democracy

Significantly, the development of telegraphic and telephonic infrastructures found its impetus in the aftermath of the 1847 Sonderbund War, the

22 «Le Conseil fédéral n›a pas hésité un instant à admettre qu›on doit comprendre sous la désignation "de télégraphes électriques" toutes les installations destinées à échanger, au moyen de l'électricité, des pensées entre deux points plus ou moins éloignés». FF, 4/55, 14 Décembre 1878, p. 5.

23 FF, 17/56, 22 Novembre 1884, p. 20. This overview of the origins of the telephone is based on: Direction Générale des PTT (publ. par), *Un siècle de télécommunications en Suisse, 1852–1952. Tome II, Téléphone, sources de courant et installations d'énergie, lignes aériennes et lignes souterraines* (Berne, 1959), 55–98, 98–220. On the history of the telephone in Switzerland see also: M. Hofmann, "The Telephone and Its Significance to Switzerland's First Subscribers," *Bulletin Technique PTT* 10 (1980), pp. 390–392; C. Kobelt, "100 years of Telephone Service in Switzerland," *Bulletin Technique PTT* 10 (1980), pp. 344–363.

Civil War that opposed Catholic, mostly rural and conservative cantons, to Protestant, urban and liberal ones[24]. The victory of the latter allowed a Constitution to be approved in 1848, and initiated the transformation of Switzerland into a modern state run by a managerial class capable of meeting the challenges of an industrial society coming into being.

Two key components of the new institutional set-up were the introduction of democratic features – particularly important given the period – together with a federalist structure. This meant Switzerland followed a similar path to the States, passing from being a league of practically independent states to a federal organisation with a central government[25]. The new institutional framework had actually been a root cause of the Sonderbund War, which had broken out when some conservative cantons refused to adhere to plans for centralising power put forward by the Diet.

The almost absolute power exercised between 1848 and 1914 by the Radical Liberal Party had profound consequences on the ways of running public affairs and that included telecommunications[26]. First of all, one of the mainstays of the federal economic policy was the free circulation of people, capital and goods throughout the entire national territory[27], and the telegraph appeared right from the beginning an appropriate tool for encouraging such movement. Secondly, the Federal Council looked kindly on the wide diffusion of the telegraph because it would be under the control of the newborn central government, which thus could appear in the cantons' eyes an indispensable institution for the creation of national infrastructures. Under government control, in fact, the electric telegraphy could exert "a strong and healthy influence on the moral or material unity of a country"[28]. Centralization and nation building were two of the main arguments used by the supporters of the state monopoly to persuade

---

24 P. du Bois, *Le Guerre du Sonderbund. La Suisse de 1847* (Paris: Éditions Alvik, 2002).

25 E. Weibel, "La coexistence linguistique face au défi européen," *Annales de la Faculté de Droit et de Science Politique de l'université de Clermont*, 25 (1989).

26 R. Ruffieux, "La Suisse des Radicaux 1848–1914," in *Nouvelle Histoire de la Suisse et des Suisses*, dir. G. Andrey et al., vol. 3 (Payot: Lausanne, 1982–1983).

27 H. Roh, *Fédéralisme politique et décentralisation économique et industrielle.* Thèse pour le doctorat, Université de Paris, 1960, 160.

28 «une influence salutaire fortifiante sur l'unité morale ou matérielle d'un pays». FF, 3/64, 27 Décembre 1851, p. 338.

public authorities to nationalize telecommunications services. Indeed, it was feared that left in private hands, and therefore under the influence of the "material, agricultural and industrial interests of the Cantons and their parties"[29], electric telegraphy could well risk tearing the country apart:

> Now we must not forget that Switzerland contains many elements of scission which come from the physical condition of the country [...] as well as differences in language and religion and we must be careful therefore not to multiply and enlarge these factious elements. Concerning this, the supreme authority of the country would not be indifferent to the fact that in favour of the Electric Telegraphs Geneva and Baie, for example, are closer to Paris than the Federal City [Bern] is, that St Gaul is closer to Munich than Zurich and Coire, Zurich closer to Frankfurt than the central cantons etc., given that with a similar mode of procedure, all these places and regions would link with abroad thanks to thousands of intellectual and material novelties, whereas they will have to be foreigners to other parts of the country.[30]

Indeed, around the turn of the century the Swiss Parliament identified in telecommunications an efficacious instrument for cementing national unity in a country that had only recently been formed and still contained strong potentially destabilizing centrifugal forces[31].

---

29 «intérêts matériels, agricoles et industriels des Cantons et de leurs parties». *Ibidem.*

30 «Or nous ne devons pas oublier que la Suisse porte en elle beaucoup d'éléments de scission qui tiennent aux conditions physiques du pays [...] ainsi qu'à la différence de langage et de religion; nous devons donc bien nous garder de multiplier et de grossir ces éléments d'une manière factice. A cet égard, il ne saurait être indifférent à l'autorité suprême du pays qu'à la faveur des Télégraphes électriques, Genève et Baie p. ex. se trouvassent plus rapprochés de Paris que ne le serait la ville fédérale, que St-Gall fût plus rapproché de Munich que Zurich et Coire, Zurich plus rapproché de Francfort que les Cantons du centre, etc., attendu que par un pareil mode de procéder, toutes ces localités et contrées se rattacheraient à l'étranger par mille nouveaux liens intellectuels et matériels, tandis qu'ils devraient demeurer étrangers aux autres parties du pays». *Ibidem.*

31 For its capacity to set up a network of correspondences and a unifying grid between the various federal states, the electric telegraph was considered a determining means also in other countries. For the United States see R. John, *Network Nation*, p. 108 and for Italy S. Fari, G. Balbi and G. Richeri, "History and Historiography of Telecommunications in Italy," *History of Technology*, forthcoming. This centripetal function of telecommunications originated with the optic telegraph, the first point-to-point system invented by Claude Chappe in France. Chappe himself

A further significant innovation introduced by the 1848 Constitution was the establishment of universal male suffrage on both federal and canton levels. Though the application of the principle was at times still ambiguous[32], the reform showed the ideological change the liberal-radical ruling classes intended to imprint on the country. The democratic process under way could not however be limited to the state institutions alone, but needed to involve other spheres of social life, including access to telecommunications.

The structure of the telegraph network, for example, had to keep this well in mind, as the government recalled in June 1867:

> In the great monarchies of Europe the tendency prevails to link by telegraph the most important points in the country. In a small democratic state however, *where all places have the same rights according to the constitution*, and where there is the habit to have the same regard for small communities as big cities, *the country's telegraph network must satisfy this democratic feeling*. In brief, the network must be established so that the smallest community can have a telegraph bureau.[33]

---

clarified in 1793 that his telegraph was the best answer to those who thought that France was too widespread a nation to form a republic. It allowed an immense population to be reunited in one point, so turning out to be a determining fact in the moulding of a country into a nation state, see D. Headrick, *When Information Came of Age. Technologies of Knowledge in the Age of Reason and Revolution, 1700–1850* (London & New York: Oxford University Press, 2000), p. 198; on the optical telegraph see also P. Flichy, *Une histoire de la communication moderne: espace public et vie privée* (Paris: La Decouverte, 1991).

32 Over the gaps in the application of democratic principles, see also the study on the results of the 1853 Federal elections in G. Kreis, *Der Weg zur Gegenwart. Die Schweiz im neunzehnten Jahrhundert* (Basel: Basel Birkhäuser, 1986).

33 «dans les grands Etats monarchiques de l'Europe prévaut la tendance qui consiste à relier entr'eux, au moyen des télégraphes, les points les plus importants du pays. Dans un petit Etat démocratique cependant, *où toutes les localités ont les mêmes droits en vertu de la constitution elle-même*, et où l'on a l'habitude d'avoir autant d'égards pour les petites communautés que pour les grandes villes, *le réseau télégraphique du pays doit donner satisfaction à cet esprit démocratique*. En un mot, le réseau doit donc être établi de manière que la commune la plus infime puisse se procurer un bureau de télégraphes». FF, 2/28, 29 Juin 1867, p. 230, [*our italics*].

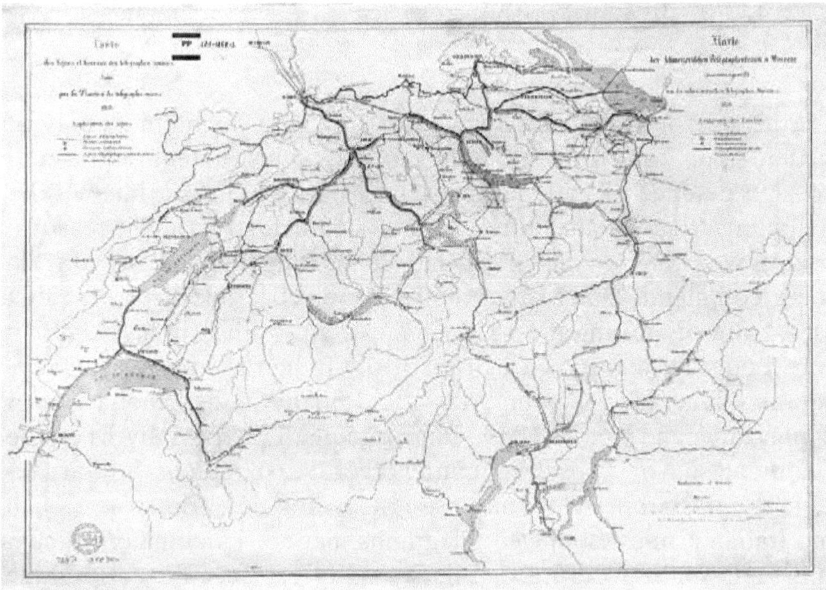

Figure 1: Map of the Swiss Telegraph Network in 1858.

Source: Museum of Communication, Bern.

The task of the Swiss telegraphic network was therefore not only to connect the principal cities in the country, but also guarantee equal access to all communities. The right to have and duty to provide the same access to communications networks, known today as a "universal service", was by no means a generalized concept at the end of the nineteenth century[34]. Yet the Swiss government highlighted right from the beginning the need to put all citizens the same level and, according to the scholar Alex Fischer, this interpretation of the concept of universal service is one of the major features of the Swiss approach to communications still in force today[35].

34 On the historic evolution of the concept of "universal service" and modifications in the course of history, see H. S. Dordick, "The origins of the universal services: History as a Determinant of Telecommunications Policy," *Telecommunications Policy* 14 (1990).

35 A. Fischer, "Swiss telecommunications policy: From state monopoly to intense regulation," *Flux* 2–3/72 (2008), 79–80.

1.3 Neutrality and defence

Another reason that made Switzerland favour a public monopoly of the telegraph was its potential relevance as an instrument for the control and defence of national territory, especially in wartime[36]. Telecommunications at that time were strategic for two main reasons: linking divisions on the front and connecting the forces with the home command. During the 1850s, the electric telegraph was first used militarily during the 1854 Crimean War[37], then in the Algerian War fought by the French in 1856[38], the Indian Mutiny War involving the British the following year[39], and finally in the Second War of Italian Independence of 1859. During the latter, especially when the Piedmontese Army was overrunning other Italian regions, military telegraphy was properly put into use for the first time. Both the French and Italian armies employed telegraphs, batteries and materials custom built for use during conflicts calling for specific non-routine equipment. Many of the techniques experimented in the Italian Wars were indeed re-used, improved and extended during the American Civil War (1861–6), which is often remembered as the first occasion in which telecommunications were deployed for military purposes on a wide scale[40].

---

36 J. Foreman-Peck, "L'état et le développement du réseau de télécommunications en Europe à ses débuts," *Histoire, économie et société*, 3 (1989), 383–384.

37 D. R. Headrick, *The invisible weapon. Telecommunications and international politics 1851–1945* (New York: Oxford University Press, 1991), 17.

38 P. Laurencin, *Le Télégraphe, terrestre sous-marin, pneumatique* (Paris: J. Rothshild Editeur, 1877), 349.

39 E. D'Amico, *Cenni sull'amministrazione dei telegrafi in Italia dalle origini all'anno 1885* (Roma: Tipografia Cecchini, 1886), 124.

40 C. Matteucci, *Manuale di telegrafia elettrica* (Torino: Unione Tipografico Editrice, 1861), 370–371; A. Suarez Saavedra, Tratado de Telegrafia, I, *Historia Universal de la Telegrafia* (Barcellona: Imprenta de Jaime Jepùs, 1880), 608–610; E. Blavier, *Nuovo trattato di telegrafia elettrica. Corso teorico e pratico ad uso dei Funzionari dell'amministrazione telegrafica, degli ingegneri costruttori, inventori, impiegati delle ferrovie* (Livorno: P. Vannini e figlio, 1874).

Since the fifties, many countries had actually been thinking how to use the telegraph in order to organise their national defence[41] and the Swiss paid close attention to military use of telecommunications from very early on. Already in 1853 Matthäus Hipp (one of the major contributors to the introduction of the telegraph in Switzerland, later head of the Bern Telegraph Workshop) presented the Federal Council with a report on the military uses of the new instrument and two years later the first exclusively military telegraph line was built between Thun and Bern. Then in 1868, Louis Curchod (head of Swiss Telegraphs and later of the Bureau) presented an organization project for a wartime telegraphic network distinct from civilian ones. In 1877, the first early experiments of telephony in the army were carried out and in the first decade of the new century the Swiss Army was one of the most active in experimenting new radiotelegraphy systems. Many of the key figures running Swiss Telecommunications came in fact from military circles. Curchod himself, as well as Karl Lendi and Emil Frey (all heads of Swiss Telegraphs and then of the Bureau), had served as high-ranking Army officers[42].

The Federal Council held that the telegraph was indispensable in the case of war in order to organize a rapid reaction for the country's defence. It was indeed the political existence of Switzerland, or as recorded in another quotation, "the very independence of our country"[43] which was at stake:

> In times of war or imminent danger coming from abroad, there would also be the great advantage of being at once informed besides a communication in time of events that interest us, a prompt mobilisation of troops. A speedy movement of the troop can influence decisively not only our financial relations but still greater the political existence of our country.[44]

41 The Italian case is discussed in detail in S. Fari, "Uccide più la parola che la spada. Telecomunicazioni e questioni militari nell'Italia del XIX secolo", *Ricerche Storiche* 1 (2006).

42 Direction générale des PTT (publ. par), *Un siècle de télécommunications en Suisse, 1852–1952. Tome III, Téléphone (suite et fin)* (Berne, 1963), 385–424 and <http://en.academic.ru/dic.nsf/enwiki/151040>, last access 3 October 2013.

43 «la indépendance même de notre patrie». FF, 3/64, 27 Décembre 1851, p. 337.

44 «Dans les temps de guerre ou de danger imminent et provenant du dehors, il y aura aussi pour eux sans doute le grand avantage d'être promptement informés

Quoting the very existence and political independence of the country was akin to another of Switzerland's founding concepts – neutrality[45]. According to some authors, in fact, during the nineteenth century the two concepts were interchangeable, given that if Switzerland had given up its neutrality, it would have lost its independence. In other terms, Swiss neutrality, backed and promoted by heavyweights like Great Britain, was the only way such a small country could affirm and protect its identity within Europe[46].

Switzerland had travelled far to reach its statute of neutrality. Already with the 1648 Westphalia Treaty the Swiss Cantons had managed to keep aloof from the conflict, obtain recognition of their independence from the Holy Roman Empire and play an important mediatory role in the peace process. However, it was much later during the 1815 Vienna Congress in fact, that the main European powers granted Switzerland the right to perpetual neutrality[47].

The statute of neutrality meant that neither then nor now the absence of a strong well-organized army – on the contrary, paradoxically, quite the opposite[48]. Even more in the course of the second half

---

d'ailleurs une communication en temps utile des événements qui nous intéressent, une prompte mise sur pied de troupes, un ordre accéléré de marches de troupes peut être d'une influence décisive non pas seulement pour nos rapports financiers, mais encore pour *l'existence politique de notre pays*», FF, 3/62, 15 Décembre 1851, pp. 293–294, [*our italics*].

45 On the history of the concept of neutrality and its importance for the development of Switzerland see H. U. Jost, "Origines, interprétations et usages de la «neutralité helvétique," *Matériaux pour l'histoire de notre temps* 93 (2009). For an international and European dimension of the same topic see M. Liniger-Goumaz, *La Suisse, sa neutralité et l'Europe* (Genève: Éditions du Temps, 1964). On neutrality and international organizations, see G. Perrin, *La neutralité permanente de la Suisse et les organisations internationales* (Heule: UGA, 1964), where however the focus is on the EU and the UN.

46 A. G. Imlah, *Britain and Switzerland 1845–60: a study of Anglo-Swiss relations during some critical years for Swiss neutrality* (Hamden, Connecticut: Archon Books, 1966).

47 A. Morin, *Les lois relatives à la guerre selon le droit des gens moderne, le droit public et le droit criminel des pays civilisés* (Cosse: Marchal & Billard, 1872).

48 See M. Milivojević & P. Maurer (eds.), *Swiss neutrality and security: armed forces, national defence, and foreign policy* (New York: Berg and St. Martin's Press, 1990).

of the nineteenth century, where Switzerland found itself surrounded by countries torn apart by internal conflicts, it needed a "strong army" organized so as to mobilize troops quickly for self-protection. Consequently, the telegraph with its inherent capacity to guarantee a quick exchange of information was soon considered a formidable defence weapon. An illuminating document dated 1895 shows in fact the military importance the Federal Council attributed to an efficient system of communication:

> Since 1888 the Federal Bureau of the general staff has signalled on many occasions the lacuna in the Swiss telegraph network consisting in the absence of direct telegraph communication between Andermatt-Dissentis, Dissentis-Olivome and Hôtel Grimsel-Innertkirchen. The installation of these three communications was insistently called for, because of their *very great importance for the defence of the country* [...]. Of these lines the most important are Gletsch-Andermatt and Dissentis-Andermatt because they guarantee telegraph communications between the *Swiss interior* and the cantons of Valais and Graubünden. In the case of detachments of our troops operating in Haut-Valais or in the Oberland, it is most important that these troops' communications via the Gotthard with the commander of the army and the Swiss inland be better than they are at the moment, as they are very vulnerable to destruction in the case of aggression near Martigny or Coire and Sargans.[49]

49 «Dès 1888 le bureau fédéral de l'état-major général a signalé à maintes reprises la lacune que constitue dans le réseau des télégraphes suisses l'absence de communications télégraphiques directes entre Andermatt-Dissentis, Dissentis-Olivone et Hôtel Grimsel Innertkirchen. L'installation de ces trois communications était demandée avec insistance, à cause de leur *très grande importance pour la défense du pays.* [...] De ces lignes, celles de Gletsch-Andermatt et Dissentis-Andermatt sont les plus importantes, puisqu'elles assurent les communications télégraphiques entre *l'intérieur de la Suisse* et les cantons du Valais et des Grisons. Dans le cas où des détachements de notre armée opéreraient dans le Haut-Valais ou dans l'Oberland des Grisons, il est de la plus haute importance que les communications de ces troupes par le Gothard avec le commandant de l'armée et avec l'intérieur de la Suisse soient mieux assurées qu'elles ne le sont actuellement, étant très exposées à être détruites en cas d'agression du côté de Martigny ou de Coire et Sargans». FF, 3/40, 11 Septembre 1895, pp. 62–65, [*our italics*].

Among the army staff and the government, the existence of telegraph lines along the Alps was judged crucial for the country's defence. It fitted in with the fact that in the same period the idea of the National Redoubt, so famous during WWII, was beginning to be formulated. Colonel Max Alphons Pfyffer von Altishofen drew up the Gotthard fortification project with an integrated efficient communications system, started in 1886[50]. The strategic relevance of the telegraph was understood also by the Ministry of Defence, which saw telecommunications as instrument contributing towards making the country safer and protecting it from any possible attack[51].

1.4 Geography and international relations

Telegraph lines were indeed cheap infrastructures, with their simple poles holding up the wires allowing the transmission of electric impulses; furthermore, for easier maintenance they often followed roads and railways. Cheap does not, however, mean easy to set up: indeed telegraph networks like land telecommunications in general are often very much affected by the lie of the land. For example in France, where most of the land is flat or not particularly hilly and where the centre has always held tight control over the peripheral areas, a radial telegraphic network developed, where Paris was linked to all the main cities but the latter had no direct connections to one other. Differently, in Italy, cut off from the rest of Europe by the mountainous chain of the Alps and sliced in half from north to south by the Apennines, the national telegraph network spread widely first in the Po Valley before

50 H. Rapold, *Strategische Probleme der schweizerischen Landesverteidigung im 19. Jahrhundert* (Frauenfeld: Huber, 1951). See also the entry, Pfyffer, Max Alphons (von Altishofen) in the *Historical Dictionary of Switzerland*, <http://www.hls-dhs-dss.ch/textes/i/I24142.php>, last access 3 October 2013.

51 V. Schneider, *Die Transformation der Telekommunikation. Vom Staatsmonopol zum globalen Markt, 1800–2000* (Frankfurt am Main: Campus, 2001).

extending down the two great coastal lines, the Tyrrhenian and the Adriatic.

It follows that in Switzerland, too, the physical geography of the land influenced political decisions. The country had two indisputable features with a determining weight both on the development of communications and its political-economic logics[52]. Firstly, the mountains had always made internal communications and transport difficult and then a central position in Europe meant sharing borders with some of the heavy-weight nations of the period.

As Géraldine Pflieger writes, the Swiss government saw the realization of transport, energy and communications networks as providing one of the country's few possibilities "of ridding itself of [the] physical determinism"[53] imposed by the mountains. Transport and telecommunications were indeed considered ways of overcoming logistic limits created by the morphology of the territory.

Then, being sited at the centre of Europe meant Switzerland was in a strategic geographical position for at least two reasons. First of all, it was the main hub for land communications, which meant that very often two states desiring to communicate with each other had no alternative but to send their telegraph and telephone messages via Switzerland. Furthermore, because of its position, the country was favoured in more narrowly political-diplomatic terms: for example, more easily than others Switzerland had a natural bent for carrying out the role of bonding agent between the different European countries and consequently also furthering the circulation of information[54].

For all these reasons Switzerland had a very high interest in linking up its telegraph networks with foreign ones. In 1851, before a domestic system had been set up, the Federal Council already declared that the country could not afford to be excluded from the "communication fever" which appeared to be invading Europe:

52 A.-L. Sanguin, *La Suisse, essai de géographie politique* (Gap: Editions Ophrys, 1983).

53 «possibilité d'affranchissement du déterminisme physique», see G. Pflieger, "La Suisse est-elle un espace première classe?," *Flux* 2–3/72 (2008), 5.

54 PTT (publ. par), *Un siècle de télécommunications en Suisse. Tome I,* 286.

> There is no direction in which in our days Europe has gone more powerfully that in the ways of facilitating the circulation and acceleration of communications. Railways [...] But still more remarkable is the speed with which the communication of thought has taken place through the telegraph.[55]

In order not to be cut out of the picture, Switzerland started both constructing a domestic network and establishing conventions with bordering and non-bordering countries. Between 1852 and 1856, it signed treaties for regulating telegraph communications with Austria, Belgium, France, Prussia, Piedmont and the Grand Duchy of Württemberg.

It was indeed its geographical position together with its special diplomatic statute of perpetual neutrality which post 1863 allowed Henri Dunant and his colleagues to found and run the Red Cross in Geneva. The same also led Switzerland to take on the managing role in the Telegraph Union (and later the International Telecommunication Union) from 1868, a position it held practically up to WWII.

1.5 Liberalism and telecommunications

The party that won the Sonderbund War had already represented the majority in the Federal Diet from the 1830s onwards, making Switzerland together with Great Britain and Belgium one of the first European states to have a liberal structure.

From the 1850s, liberalism became the main economic paradigm in Europe and as such was adopted by almost all governments in the period. As is well known, the liberal doctrine had found its first theorization in Adam Smith's *The Wealth of Nations* published in London in 1776, a reaction to the seventeenth and eighteenth century mercantile system

55 «Il n'est aucune direction dans laquelle l'Europe ait pris de nos jours plus puissante que dans les moyens de faciliter la circulation, l'accélération des communications. Les chemins de fer [...] Mais plus remarquable encore est la rapidité avec laquelle la communication des pensées a lieu au moyen des Télégraphes». FF, 3/62, 15 Décembre 1851, pp. 292–293.

which had stimulated strong government intervention in encouraging exports and discouraging imports. According to the mercantile vision, in fact, the state had to accumulate precious metals, bullion and coins, and to do this exports had per force to be higher than imports. Smith claimed, on the contrary, that enterprises had to prosper without state intervention, that duties should be all abolished, and that it would be the market with its balancing mechanism between demand and offer which would determine the growth or demise of an enterprise and therefore of a nation. The theoretical tenets of liberalism and *laissez-faire* were then re-enforced by David Ricardo and other British economists during the nineteenth century, and found direct application in the economic policies of the liberal governments. The mainstays of the *laissez-faire* doctrine and the resulting economic policies remained however those identified by Smith in his times: 1) support to free competition among enterprises; 2) absolute faith in the market's self-regulating mechanisms (the so-called "invisible hand"); 3) scarce state intervention in economic affairs; 4) absence of protectionist measures.

In line with these principles, in the period running up to the approval of the Swiss Constitution, numerous institutions that had characterized the post-Napoleonic restoration were abolished, mainly the remaining feudal privileges, crafts corporations and bourgeois hereditary rights[56]. From 1848 onwards, the new constitutional structure then allowed the application of economic principles clearly inspired by *laissez-faire.*

For example, the Federal Administration was kept very small, limited to fifty or so officials[57]. Consequently, many sectors including strategic ones like the railways were left to the initiative of private enterprise[58]. In the same period, the Swiss government was persuaded that enterprises had to be given the possibility to compete freely in the European market and new forms of communication, like the telegraph, could benefit commercial activity:

---

56 C. Gilliard, *Histoire de la Suisse* (Paris: PUF, 1987).

57 Ruffieux, "La Suisse des Radicaux".

58 Kreis, *Der Weg zur Gegenwart.*

> As for the advantages for business transaction in general, we attach to affairs in general a far higher price given that a great part of *our Swiss population* is, because of the effects of the circumstances and culture of our land, *confined to commerce and industry*. All branches of commerce and industries are more or less hindered by competition. In the middle of this incessant struggle, of businessmen's assiduous attempts tending to take away from others who work in the same branch of commerce or industry, the most favoured will be those who are better informed than others about the events concerning their branch, who will know the real price of goods in the grand markets, who will be able to satisfy promptly the demand, who will be able to carry out quickly commands and commissions and exploit them.[59]

The government's main worry was that Swiss companies might run the risk of not being able to withstand foreign competition[60] and the risk was greater still without the rapid systems of communication that other economically more advanced nations had already adopted.

In appearance, the choice to run telegraph communications as a public monopoly was an explicit contradiction of the basic principles of *laissez-faire*. Nevertheless, it found some interesting theoretical justifications in indisputable defenders of liberalist measures. For example, in 1869 the British Parliament decided to nationalise their telegraph system and entrust it to the Post Office. As in the period the country was the cradle of free-trade, theoretical explanations were needed for allowing such a measure not to jar with the tenet of scarce public intervention. What was the innermost reason which brought Gladstone's liberals, economists like

---

59 «Quant aux avantages, résultant pour les relations d'affaires en général, nous y attachons un prix d'autant plus haut qu'une grande partie de *notre population* suisse est, par un effet des circonstances de culture du notre sol, *réduite au commerce et à l'industrie*. Toutes les branches de commerce, toutes les industries se trouvent plus ou moins gênées par la concurrence. Au milieu de cette lutte incessante, de ces efforts assidus des hommes d'affaires, tendant à l'emporter sur les autres qui exploitent la même branche de commerce ou d'industrie, celui-là sera le plus favorisé qui sera informé plus tôt que d'autres des événements réagissant sur la branche qui le concerne, qui sera au fait du prix réel des marchandises sur les grands marchés, qui pourra satisfaire promptement aux demandes, qui sera à même d'exécuter avec célérité les commandes et commissions relatives à son exploitation». FF, 3/62, 15 Décembre 1851, p. 294, [*our italics*].

60 FF, 3/64, 27 Décembre 1851, p. 337.

William Stanley Jevons and Alfred Marshall and business men like John Lewis Ricardo to support the nationalization of the telegraph service against the interest of private companies? Why was it that the defenders of free competition did not spend a word in defence of the telegraph companies and on the contrary backed the process that would prevent them from carrying on with their activity? Charles Richard Perry gives a possible answer: "Free trade in a strongly capitalist system and government intervention in that same system were not mutually exclusive ideals"[61]. He goes on to say: "extensive institutional reform was perfectly compatible with a commitment to individualism and market forces, and was indeed a prerequisite of a competitive society"[62]. This concept is better understood after examining the specific case of John Lewis Ricardo, nephew of David, chairman of the *Electric Telegraph Company* from its beginnings until 1857 and also a member of Parliament. An exemplary defender of competition, he fought strenuously in parliament against laws designed to set limits to such freedom, like the Corn Laws and the Navigation Acts. Nevertheless, despite both his personal experience with one of the most important telegraph companies and his parliamentary career, he sent Gladstone (then Chancellor of the Exchequer) a memorandum "in support of the expediency of the telegraphic communications of the United Kingdom being in the hands of HM Government and administered by the Post Office"[63]. He wrote that the running of the telegraph service needed to be in the hands of the public administration, following the example of many other European countries. Probably, as Perry holds, Ricardo thought that a reliable public telegraph network would be one of the best institutions for supporting free competition, and this was why he backed it in apparent contradiction with his ideas of political economy.

> Moreover, if one examines Ricardo's analysis of the shipping industry under the Navigation acts and his description of the telegraphs before nationalisation, the parallels are remarkable. In each an oligopolistic system caused unnecessarily high prices and inhibited the expansion of the economy. Only his solutions

61 Perry, *The Victorian Post Office*, 91.

62 *Ibidem.*

63 Memorandum of John Lewis Ricardo quoted in Kieve, *The Electric Telegraph*, 120.

> differed. In the case of the shipping industry, free trade was the remedy; for the telegraphs, nationalisation.[64]

Following this idea, the public running of the Swiss telegraphs would be in line with the principles of liberalism if telegraph communications were seen as an institutional expedient for guaranteeing free competition rather than purely as a market. Furthermore, always in order to favour free competition, the Federal Council tried to cut down to a minimum both national and international telegraph charges. The domestic telegraph tariffs introduced in Switzerland in 1851 were the same as in France and Germany, but the Federal Council made a series of moves to reduce them. According to the government indeed "A first glance at the [German] tariff shows us that if we want to obtain our dual aim, using the organisation for the public and collecting net profits, we must lower the tariffs considerably"[65]. On the other hand, the Federal Council realized at once that a reduction in charges would encourage greater use and, therefore, a need to build new networks[66]. It was Parliament however – again with a liberal majority – which put up most opposition against reducing charges for fear of increasing the public debt[67]. Above all Parliament declared in 1877 that the telegraph service was not yet commonly used in Switzerland, so that making the whole population

64 Perry, *The Victorian Post Office*, 91.

65 «Un premier coup d'œil sur ce tarif nous fait voir que si nous voulons atteindre notre double but, savoir d'utiliser l'institution pour le public et percevoir le produit net, nous devons faire subir aux taxes une réduction considérable». FF, 3/62, 15 Décembre 1851, p. 292.

66 «La réduction des taxes des télégrammes aura certainement une double influence, savoir, d'un côté, d'augmenter d'une manière notable le nombre des dépêches expédiées par les bureaux actuels, d'un autre côté, de faire plus vivement sentir le besoin de nouvelles lignes et de nouveaux bureaux. Ainsi la réduction des taxes aura pour conséquence d'obliger à modifier le réseau des télégraphes suisses» ("The result of the reduction in telegram charges will certainly have a dual influence, on one hand knowing how to increase considerably the number of telegrams sent from existing offices, on the other making the need for new lines and new offices felt. So the reduction in charges will create the need to modify the Swiss telegraph network"). FF, 2/28, 29 Juin 1867, p. 229.

67 FF, 1/11, 17 Mars 1877, p. 422.

uphold the expense for a further drop in charges would be contrary to the democratic nature of the country:

> The result of the research done by the telegraph administration that only 3% to 4% of the population use the telegraph. It would be unfair therefore to make the whole population bear the expense in a more or less equal proportion, since the benefits of this administration only go to a small number of people, who are comfortable off, in favourable circumstances. It would however be like this, if the State coffers had to sustain the expenses of this branch of the administration.[68]

In spite of this stand-off lasting right through the nineteenth century, Swiss telegraph tariffs were constantly and markedly lower than elsewhere, as shown by the fact that in 1865 Switzerland had the cheapest domestic charges in Europe, and these were to remain unchanged from 1877 to 1917[69]. Swiss telegraph tariffs were indeed an exception, for while most other European countries chose to adopt expensive rates in order to limit use to certain categories and social classes, Switzerland aimed right from the beginning to make the telegraph a service "within everybody's reach and at the service of the country's economy"[70].

There is another important element for understanding the specific Swiss treatment of telegraph charges and interest in international regulations. Switzerland was able to keep its domestic charges so low because of its position in the centre of Europe, which meant it made considerable profits from international transit traffic, i.e. the small percentages collected when a communication between two foreign countries

68 «le résulte de recherches faites par l'administration des télégraphes que 3 à 4% de la population seulement fait usage du télégraphe…Il serait donc injuste de faire supporter les frais du télégraphe a toute la population dans une proportion à peu près égale, alors que les bienfaits de cette administration ne profitent qu'à un petit nombre de personnes, se trouvant pour la plupart dans une position aisée et des circonstances favorables; il en serait cependant ainsi, si la caisse de l'Etat devait supporter les frais de cette branche de l'administration». *Ibidem.*

69 For Swiss telegraph charges see PTT (publ. par), *Un siècle de télécommunications en Suisse, Tome III*, 543–549.

70 «à la portée de chacun et au service de l'économie du pays», see S. Pravato, "De Télécom PTT à Swisscom," in *Du monopole à la concurrence. Analyse critique de l'évolution de 6 entreprises suisses*, dir. M. Finger, S. Pravato, J.-N. Rey (Lausanne: LEP, 1994), 71.

transited through the Swiss networks on the way to destination. And, as Parliament declared, domestic telegraph traffic was running at a clear loss, though "this loss has been covered so far by the profit coming from the higher revenue from international and transit telegrams and municipality services and various other administration returns"[71]. With its domestic traffic alone Switzerland would in fact have been incurring heavy losses, so that the country was able to allow itself the lowest telegraphic charges in Europe only because it obtained high returns from international traffic. It is not surprising therefore that the Federal Council was so heedful to international telegraphy and watchful of its developments.

1.6 Foreign trade

The economic liberalism that lay behind the actions of the Swiss political class was signalled from 1848 onwards by a series of measures aiming at favouring free exchange, like the abolition of customs duties and joining the Latin Monetary Union (1856). The expansive effects of this policy meant that the Swiss economy, which lacked natural resources but was well equipped with qualified technicians and abundant capitals, could steer towards exportation[72]. Obvious, therefore, that the development of telecommunications was backed and in some cases demanded by civil society and in particular the business classes. A famous example is the letter that on 22 April 1851 Zurich's 18 leading commercial companies sent to the Federal Council explicitly requesting the construction of an electric telegraph; they were worried about possible damage from a drop in foreign trade:

71 «Cependant cette perte a été couverte jusqu›à présent par le bénéfice réalisé sur le produit plus élevé des télégrammes internationaux et de transit, par les prestations des communes et par diverses autres recettes de l'administration». FF, 1/11, 17 Mars 1877, p. 423.

72 Ruffieux, "La Suisse des Radicaux".

> Switzerland, a country partly dedicated to commerce and industry, keeps up relations of all kinds with states in all continents. Now, already now several neighbouring countries, notably Austria, Württemberg, the Grand Duchy of Baden and France have set up telegraph links close up to our borders. Switzerland must go with progress and extend these lines into its territory, if it does not want to be cut out of external movement. This motive alone justifies the pressing need, expressed everywhere in our country, for the telegraph to be introduced into Switzerland.[73]

The telecommunications networks were therefore also indispensable for Switzerland's international economic activities. Beyond the political reasons analysed above, Switzerland clearly based its economic well-being on foreign trade, as the Federal Council notes in an enlightening intervention in favour of extending the railway network:

> In the whole of the European continent there is perhaps no country which is so little to itself for extracting from the ground the essential products for life, and which consequently has more interest in obtaining cheaply its consumer products, raw, and export without high costs its manufactured goods. There is perhaps no other country which to maintain its well-being has to care about the interests of its industry and the speedy transport of persons and goods; no country to which the English proverb "Time is money" can be so truthfully applied.[74]

---

73 «La Suisse, pays voué en parti au commerce et à l'industrie, entretient des relations de diverse nature avec des Etats de tous les continents. Or, déjà maintenant, plusieurs pays voisins, notamment l'Autriche le Wurtemberg, le Grand-duché de Bade et la France, ont établi des liaisons télégraphique jusqu'à proximité de nos frontières. La Suisse se doit de marcher avec le progrès et de prolonger ces lignes sur son territoire, si elle ne veut pas être coupée des marches extérieures. Ce Motif à lui seul justifie déjà la demande toujours plus pressante, formulée de toute part dans notre pays, d'introduire le télégraphe en Suisse». PTT (publ. par), *Un siècle de télécommunications en Suisse. Tome I*, 118.

74 «Dans tout le continent européen, il n'y a peut-être pas un pays qui soit si peu à même que la Suisse de tirer du sol les produits indispensables à la vie, et qui par conséquent ait plus d'intérêt à obtenir à bon compte ses objets de consommation, ses matières premières, et à expédier à l'étranger et sans beaucoup de frais ses articles de fabrication. *Il n'est peut-être pas de pays qui, pour conserver son bien-être, doive autant avoir à cœur les intérêts de son industrie et la rapidité du transport des personnes et des marchandises*; pas de pays auquel on puisse appliquer avec autant de vérité le proverbe anglais que le temps vaut de l'or». FF, 3/64, 27 Décembre 1851, pp. 344–345, [*our italics*].

For telecommunications too, establishing bilateral relations and being at the heart of a European network would therefore turn out to be an economic advantage for the country.

1.7 Radicalism and telecommunications

Victory in the Sonderbund War had indeed allowed the Swiss middle classes to extend their power from economics to politics. The ideology permeating the new political class's choices was a form of Radicalism, a current of thought that brought about a democratizing process in society well ahead of the rest of continental Europe. In concrete terms, Radicalism favoured strengthening the State's role, without reducing freedom of enterprise. State action was seen as promoting social mobility by secularizing the institutions, extending rights for the populace and creating a State school system capable of guaranteeing equal access to education for all social classes[75]. This doctrine was not only extended and democratized by telecommunication networks, but also underlay the lowering of telegraph and telephone charges. Keeping rates down opened up the new services to a far wider section of the population, a policy which had already been sustained back in 1853, by the then Head of Swiss Telegraphs, Karl Brunner, as the following quotation shows: "Switzerland was the first country to introduce a uniform and very low tariff for all distances and try to popularize as far as possible the use of the telegraphs [...] an instrument created by the authorities to facilitate public relations"[76]. It was one of the aims that the Federal Council held a State like Switzerland had to pursue:

---

75 A. Tanner, "Das Recht auf Revolution," in *Im Zeichen der Revolution: der Weg zum schweizerischen Bundesstaat, 1798–1848*, ed. T. Hildbrand and A. Tanner (Zürich: Chronos, 1997).

76 «La Suisse est le première pays qui, en introduisant un taxe uniforme et très minime pour toutes les distances, ait essayé de *vulgariser autant que possible l'usage des télégraphes* [...] instrument créé par les autorités pour faciliter les relations des public», see PTT (publ. par), *Un siècle de télécommunications en Suisse. Tome I,* 248–249, [*our italics*].

> Because of the drop in charges, the telegraph will daily become a property common to the whole world and will enter the habits of all families just as the post does now. We can even forecast with surety that in 20 years' time telegraphs *will have become such a habit that they will seem to be an absolute necessity like schools, post offices, hotel and shops.*[77]

For the Federal Council, therefore, telecommunications also covered a basic function in the social life of the country in a vision held right from the early years of the development of the telegraph, when it commented that the new means of communication "would be of a marked utility, in all cases very convenient and pleasant to use, also with family and friends and in scientific, artistic and official life"[78].

1.8 Know-how and technical elite

Finally, there were also technical reasons that can help explain the importance given to telecommunications in Switzerland.

The country had, for example, a long tradition in horology that went back to Louis XIV's 1685 Fontainebleau persecutory edict forcing Huguenots to emigrate from France to other countries. Part of that rich community moved to French-speaking Switzerland, where they set up the flourishing Swiss industry of horology[79]. A technical know-how in the sector developed side by side with the business activity. Initially

77 «Par la réduction des taxes, le télégraphe deviendra tous les jours plus une propriété commune à tout le monde et entrera dans les habitudes de chaque famille absolument de la même manière que la poste maintenant. Il est même permis de prédire avec certitude que dans 20 ans on aura tellement pris *l'habitude du télégraphe qu'il paraîtra à chacun d'une nécessité aussi absolue que l'école, le bureau de poste, l'auberge et la boutique*». FF, 2/28, 29 Juin 1867, p. 230, [*our italics*].

78 «aussi dans les relations de famille et entre amis, dans la vie scientifique, artistique et officielle sera d'une utilité signalée, et en tout cas très – commode et agréable». FF, 3/62, 15 Décembre 1851, p. 295.

79 C. M. Cipolla, *Le macchine del tempo: l'orologio e la società, 1300–1700* (Bologna: il Mulino, 1980).

the making of clocks functioned as a cottage industry, especially in the Canton of Neuchâtel and the city of Geneva, where thousands of artisans turned out clocks, precision instruments and carillons. Later, in the 1830s, they began to be taken on in specialized workshops, like Vacheron & Constantin's[80].

The horology section furnished a fundamental technical basis for the telegraph industry. Given the nature of clocks as high precision instruments with various cogwheels having to function in synchrony, they were constructed by top level technicians who had been sending down their art for centuries. Telegraph apparatus was equally complicated to make since different cogwheels had to synchronize mechanically. And this was the main reason why in Switzerland and elsewhere, clockmakers were the first to be employed in the construction of telegraphs. Furthermore, in Switzerland watchmakers and ex-watchmakers were among the principal promoters and makers of the networks. The most famous was probably Louis François Clément Breguet, of French birth but educated in Neuchâtel, the grandson of the founder of the eponymous watch manufacturing company. Together with Alphonse Foy, he was the creator of France's first electric telegraph, imitating the structure of the optic telegraph[81].

Another famous name shared between telegraphy and horology was Karl Kaiser, watchmaker and later telegraph inspector at St. Gallen, who sent the Head of the Department of the Post and Public Works (William Matthias Naeff) the letter quoted above in which the city's shopkeepers and bankers invited the government to set up a telegraph network quickly[82].

---

80 J-F. Bergier, *Wirtschaftgeschichte der Schweiz* (Zürich: Benziger Verlag, 1983).

81 Huurdeman, *The Worldwide History of Telecommunications*, 73. See also E. Breguet, *Breguet horloger depuis 1775, vie et postérité d'Abraham-Louis Breguet* (Paris: ed. Alain de Gourcuff, 1997).

82 PTT (publ. par), *Un siècle de télécommunications en Suisse. Tome I,* 115.

Figure 2: An example of a Foy & Breguet telegraph.

Source: <http://distantwriting.co.uk/instruments.html>.

Figure 3: Portrait of Karl Kaiser, watchmaker and later telegraph inspector at St. Gallen.

Source: Museum of Communication, Bern.

A third key figure was Matthäus Hipp, a German watch and precision instrument maker who was appointed head of the *Atelier fédéral des télégraphes,* a training school for telegraph operators established in 1852, over applications from other candidates who were both Swiss and watchmakers, like Meinrad Theiler and Karl Kaiser[83]. It goes without saying that among the principal tasks of the *Atelier* were the study and development of the techniques of telegraph transmission in Switzerland, the construction of telegraphic apparatus and the making of precision watches[84]. Yet again, an example of the very close ties between watch and telegraph sectors in Switzerland.

Figure 4: Matthäus Hipp, head of the Atelier Fédéral des Télégraphes in Bern from 1852 to 1860.

Source: Museum of Communication, Bern.

83 *Ibidem,* chapter 5.

84 *Ibidem*, 167.

The establishment of the *Atelier* belonged to the strategy underlying the Swiss education policy of promoting higher education in cutting-edge technology. The liberal-radical elite's general project was brought to life at Zurich's Federal Polytechnic, after an almost fifty-year-long debate. Side by side with the Federal University, whose function was to guarantee the development of Swiss society in a liberal sense, the Polytechnic had been strongly desired by the industrial and economic circles, who saw its main task as giving impulse to the new economic system developed by the bourgeoisie[85].

For the telegraphic sector, Switzerland decided to place its stake on highly qualified personnel in this new field and, as we have seen, had no qualms about recruiting them from abroad. Two foreigners, for example, were called on to realize the Swiss telegraph system starting from 1851: the Alsatian Carl August von Steinheil, who built the Austrian network, and Leo Boumgartner, an engineer who had worked on the railway and telegraph systems in Lombardy. Among the Swiss employed in the telegraph administration some had considerable technical/scientific experience. Karl Brunner von Wattenwyl, geologist and professor of physics from Bern who had travelled over much of Europe with numerous scientific contacts, was appointed the first head of the telegraph administration in Switzerland. Louis Curchod, an engineer from Canton Vaud and a Paris graduate specialized in railway networks, was appointed telegraph inspector in 1852 and then took over from Brunner in 1858. Karl Lendi, from Walcnstadt, who oversaw the works for the construction of the telegraph network in Piedmont, was one of the first heads of the telegraph offices in 1852 and then followed in the footsteps of Curchod, as did August Frey from Olten[86]. All these four highly qualified Swiss professionals played key roles in the creation of the Telegraph Union and three of them (Curchod, Lendi and Frey) were to become its head.

85 D. Gugerli, P. Kupper und D. Speich, *Die Zukunftsmaschine. Konjunkturen der ETH Zürich 1855–2005* (Zürich: Chronos Verlag, 2005).

86 For biographies, see PTT (publ. par), *Un siècle de télécommunications en Suisse. Tome I,* 130; Grossi, "Le rôle International de personnalités suisses du XIXe siècle dans le domaine des télégraphes"; PTT Archive, Fernmelde-departement (personel).

Thanks to its horology tradition, Switzerland had a great advantage in respect to other countries. This advantage, initially founded on traditional craftsmanship, had then been later increased by bringing in international and national experts for the specific purpose of constructing and developing modern telecommunications (von Steinheil and Curchod are just two examples). Furthermore, as already pointed out, Switzerland had for long time been experimenting and refining its diplomatic skills, counting always on its neutrality; consequently, at that time, the country was an ideal place for promoting an international dialogue among experts speaking similar technical languages. In other words, Switzerland seemed to have another requirement for hosting international organizations that, like the ITU, would promote the domination of technical elites over the international regulation of specific services such as the telegraph.

1.9 National and international interests

From the mid-nineteenth century onwards, the Swiss Confederation saw telecommunications as a key element in running the country at several levels.

Politically, telegraphs and telephones could help cement the national unity of a recently formed State that still contained various breakaway elements. Telecommunications were indispensable for military defence, the control of the territory and in the case of war. Then they were instruments that in the eyes of the ruling classes would help cultivate already existing international relations and create new ones.

Economically, telecommunications were essential for foreign trade, a main element in the Swiss economy because national companies exported the major part of their products to neighbouring countries. Communications were also an expression of the concept of free trade, which characterized the political-economic doctrine of the Federal Council. The main objective of the government was indeed to allow

Swiss companies to compete on the same level as foreign ones, so that the constant restraint over telegraph and telephone charges are to be read in this light. It was also the expression of a particular social vision, which recalled the basic principles of democracy; the government pledged to build widespread telegraph and telephone networks in order to extend the use of telecommunications to as much of the population as possible.

Technologically, Switzerland adopted a twofold strategy on telecommunication. On one side, it exploited a previous know-how and saw the telegraph as a natural outcome and a form of diversification from the clock and precision instrument sector in which the country had proven skills. On the other, it repeated a typical Swiss attitude to education which aimed at excelling in higher professional skills, consequently, in creating a technical elite of recognized competence capable of leading the sector almost autonomously.

A last feature of the Swiss approach to telecommunications, which will be the main object of the research in the remaining part of this volume, is the declared interest the Swiss Confederation had in the creation and management of a European network of telecommunications. Between the 1850s and the 1870s, for reasons that will be analysed in the following chapters, the Swiss government saw the domestic and international running of the telegraph service as closely related. Thus Switzerland played a major role leading up to the creation of the Telegraph Union, proposed it, worked for its approval and finally ran the most important control body of this institution, the Bureau. To take on such a leading role, a small country like Switzerland had to put into the field its political, economic, technical and social expertise, applying to international telegraphy those ideas/visions on telecommunications that had been developed on a national level.

Chapter 2
"Bringing together the two large electric currents dividing Europe" (1849–1865)

Introduction

The expansion of the first national telegraph networks was followed by the first international connections in the early fifties, some by means of undersea cables, almost exclusively laid and run by the British, who for obvious historical, geographical and industrial reasons were alone in possessing the necessary technology. The telegraph was indeed useful for maintaining political, economic and social relations with neighbouring and non-neighbouring countries with whom solid ties had been forged. While both national and international lines were of a high level of technological standardization, the same could not be said for telegraph systems used, which were based on a variety of patents and regulations. Right from the beginning therefore, the issue existed of standardising technologies, codes and uniforming the norms regulating the circulation of telegrams. A common tariff also needed to be set for telegrams travelling along international lines.

Between the late 1840s and mid 1860s the international management of the telegraph went through three distinct phases. The first saw a series of bilateral treatises, the second the creation of two unions (the Austro-German Telegraphic and the Western European), while the third led to the birth of the Telegraph Union (henceforth TU), the first supranatural organization tasked with managing communications, initially at European and then world level. In all these phases Switzerland played an active role, which on the balance was decisive in the creation of a common space for European communication.

In carrying out this important role, Switzerland may well have been advantaged by Great Britain's absence from the international telegraph scene. Given that British telegraphs were privately run, the Government could not take part either in the constitution of the two Telegraph Unions or even less the TU, to be created in 1865. And though Great Britain was still the main political and economic power, three others were beginning to emerge: France, Germany and the States. Whatever, Germany was caught up in the throes of unification, France was still mainly agricultural and artisan while the US was outside the sphere of continental politics. This was why two small dynamic industrial states like Belgium and Switzerland were able to enjoy a wide margin of diplomatic manoeuvre.

2.1 Bilateral conventions

The first international conventions for regulating telegraph traffic were signed on the initiative of Prussia as far back as 1848. The reasons behind Prussia's drive to regulating telegraphs lay in its geography, economic structure and political ambitions within the German Confederation. First of all, the Prussian territory was split into eastern and western sections, as the map below shows. To connect the main cities the telegraph wires had per force to cross foreign states, so that any linking of a national network depened on agreements with other sovereign states.

Secondly, Prussia had a strong interest in forging close relations with other states containing fragmented enclaves of German peoples beginning to claim a common national identity. They were calling for the abolition of the last vestiges of the Holy Roman Empire, which had been founded on the principle of the dynastic legitimacy of states and still did not recognise the self-determination of the peoples[1].

1 D. Blackbourn, *The Long Nineteenth Century: a History of Germany, 1780–1918* (New York: Oxford University Press, 1998).

Figure 5: A map of Prussia in the first half of the nineteenth century: one nation but a divided territory.

Source: Putzger – Historischer Weltatlas, 89. Auflage, 1965.

Lastly, the Prussian Kingdom contained areas like Silesia, which were strongly industrialised, particularly in textiles and iron/steel[2]. The economic development of these regions had had a dual effect. On one hand, it had guaranteed the technological and managerial know-how essential for regulating the infrastructures required for an efficient telegraphy. On the other, it had strengthened the role of a new managerial

2 A. Herzig, *Schlesien. Das Land und seine Geschichte in Bildern, Texten und Dokumenten* (Hamburg: Ellert & Richter Verlag, 2008).

class much interested in the development of telecommunications as an indispensable tool for improving the functioning of the free market, the driving force of the nascent industrial capitalism[3]. Consequently, for reasons both of domestic traffic and political/economic objectives, Prussia signed as many as 15 international treaties with other German states in the late 1840s[4].

On 3 October 1849, the same geographic, political and economic reasons led Prussia to sign a treaty with Austria about the installation and use of telegraphy for the exchange of international telegrams[5]. This was the first international telegraph convention designed not to connect diverse regions of the same nation, but bring together different states. The convention was drawn up on the model of preceding bilateral postal treaties (which provided an example for international telegraphy in its infancy[6]). It decreed on some of the issues which re-surfaced in many later telegraph treaties, such as regulations on traffic flow and the dynamics for the border exchange of telegrams as well as tariffs. It was only the first of a series of telegraph agreements which Prussia and Austria signed with Saxony and Bavaria in 1849/1850, and which led to the creation of the Austro-German Telegraph Union in 1850.

In Central/Western Europe, bilateral telegraph conventions began to be signed a couple of years later, with the first stipulated between France and Belgium in April 1851[7]. Between then and 1855, the year of the foundation of Western Europe's Telegraph Union, five more countries signed treaties on international communication: France, Belgium, Piedmont, Switzerland and Spain. With the exception of the latter, they were all nations with liberal-inspired regimes, which perceived all means like the telegraph as instruments of peace and prosperity able to

3 J. Habermas, *The structural transformation of the public sphere: an inquiry into a category of bourgeois society* (Cambridge MA: MIT Press, 1991).

4 Codding, *The International Telecommunication Union*, 13; Durand-Barhtez, *L'Union internazionale des télécommunications*, 28.

5 *L'Union Télégraphique Internationale (1865–1915)*, 3.

6 Laborie, *La France, l'Europe et l'ordre International des communications (1865–1959)*, 129–144.

7 Meyer, *L'Union Internationale des Télécommunications et son bureau*, 2.

favour free trade. Belgium, after the 1830 Revolution[8], and Switzerland, after the 1847 Civil War, both boasted liberal institutions and political power in the hands of the middle-class. After a short period of republicanism in France, Napoleon III's coming to power in 1848 abolished the main institutions imposed by the Restoration[9]. While in Piedmont, the Savoy monarchy was forced to accept numerous instances from the middle classes and approve the 1848 Albertine Statute[10].

In this political climate, Switzerland was very active in signing bilateral treaties, and obviously because of its strategic geographical position, stipulated them both with German-speaking nations and with those gravitating in the French-speaking orbit: with Austria 26 April 1852, France 23 December 1852, Piedmont 25 June 1853, again with France and Baden 8 August 1853, Württemberg 25 August 1854 and Spain 24 November 1854.

2.2 The Austro-German Union

The Austro-German Telegraph Union (hereafter AGTU) was established on 25 July 1850 at Dresden by Austria, Prussia, Bavaria and Saxony as founding members. In the next few years other countries in the German-speaking area joined (Württemberg, Hanover, Baden, Mecklenburg-Schwerin and the Lombardy-Veneto Region) as well as some with close relations with them (Holland, the Duchy of Modena and Parma, Tuscany and the Papal States)[11]. For certain aspects AGTU achieved in the field of communications what the *Zollverein* had already

8 N. Schiffino, *Crises politiques et démocratie en Belgique* (Paris: L'Harmattan, 2003).

9 P. Milza, *Napoléon III* (Paris: Perrin, 2006).

10 D. Mack Smith, *Modern Italy: A Political History* (New Haven: Yale University Press, 1997).

11 On the history of AGTU see J. Reindl, *Der Deutsch-Österreichische Telegraphenverein und die Entwicklung des Deutschen Telegraphenwesens 1850–1871* (Frankfurt am Main: Peter Lang, 1993).

accomplished for customs. Back during the Congress of Vienna, a German Confederation had been established to guarantee "the internal and external security of Germany and the independence and integrity of the German states" (art. 2). Besides numerous small principalities the Federation included Austria and Prussia, the two European powers around which a coalition formed of those aspiring to a *Großdeutschland* under Austrian hegemony or a *Kleindeutschland* under the Prussians[12]. These aspirations held good until the definitive affirmation of Prussia and consequent foundation of the German Empire in 1871.

Figure 6: A map of AGTU's telegraph system in 1852.

Source: <http://distantwriting.co.uk>.

12 J. Droz (dir.), *Histoire de l'Allemagne. Vol 1. La Formation de l'unité allemande, 1789–1871* (Paris: Hatier, 1970).

AGTU had been established, however, in a period in which Prussia's political clout had been notably diminished by the 1848 disorders, which thus seemed to point to the German Federation passing definitively into the Austrian orbit. The first AGTU convention would serve as a model for successive multilateral treaties, so it can be credited with laying the bases for the future international organizations aiming at regulating telecommunications. Four elements in particular would underlie most of the future European conventions and would be adopted almost without change by the TU in 1865. Firstly, the treaty concerned only international communications and each country was left free to manage its domestic telegraph service at will (art. 2). Secondly, measures were established to uniform the international running of the service: from telegraph office opening hours (art. 9) to the length of telegrams (20 word minimum and 100 word maximum, art. 12), as well as the classification into state, service and private correspondence (art. 15) and costing of the number of words and distances (art. 26)[13]. Thirdly, given that telegraphy was constantly evolving, it was decided to submit the contents of the convention to periodic revisions with conferences to be held "from time to time" (art. 40). Finally, the Dresden Convention allowed "other German governments" to join at any moment (art. 41)[14].

AGTU re-assembled in periodic conferences designed to determine and harmonise tariffs, facilitate the flow of international traffic and back the introduction of technological innovations[15]. The seats of

13 The mandatory use of Morse telegraph apparatus was established the following year, during the Vienna Conference. See P. A. Carrè, "Archeologie d'une Europe des télécommunications," *Revue Française des Télécommunications*, 70, October (1989), 75.

14 See "Traité d'Etat conclu le 25 juillet 1850 entre l'Autriche, la Prusse, la Baviere et la Saxe portant création de l'Union télégraphique austro-allemande," *Journal officiel general de l'Empire d'Austriche*, 30 September 1850.

15 Codding, *The International Telegraphic Union*, 13–29; and J. Reindl, "Partikularstaatliche Politik und Technische Dynamik: Die Drahtgebundene Telegraphie und der Deutsch-Österreichische Telegraphenverein von 1850," in *Vom Flügeltelegraphen zum Internet: Geschichte der modernen Telekommunikation*, ed. by H.-J. Teuteberg and C. Neutsch (Stuttgart: F. Steiner, 1998), 34–35.

the conferences were Vienna (1851), Berlin (1853), Munich (1855), Stuttgart (1857), The Hague (1861) and Hanover (1863). In the late 1850s and early 1860s, it supplied the model that the Telegraph Union would take over by extending its range of action and bringing into its conventions private companies like the *International* and *Electric Telegraph Company,* the *Submarine Telegraph Company* and the *Company of the Telegraph Lines of the Mediterranean Islands*. After the foundation of the German Empire, AGTU was dissolved in July 1872[16]. However, its influence already began to wane from 1865 onwards, partly because of the birth of the TU, designed to overcome the division between eastern and western Europe and partly because of the 1866 Austro-Prussian War and the end of the German Confederation.

2.3 The Western European Union

The equivalent of AGTU for the Latin block had no codified name, but was conventionally known as the Western European Telegraph Union (WETU). It was constituted on 29 December 1855 in Paris by Belgium, France, Piedmont, Spain and Switzerland, which between 1851 and 1854 had stipulated a series of bilateral treaties allowing international telegraph communications in central and southern Europe. As happened with AGTU, other states joined WETU over the following years: Portugal, Holland, the duchies of Tuscany and Modena/Parma, the Kingdom of the two Sicilies and the Papal States, as well as some of the private British companies specialising in underseas cables.

AGTU had come into being mostly for vested interests and political aims. A desire for national identity was already stirring and the union was making a first, but concrete, step towards a political unification of the German states[17]. The reasons for creating WETU

16 Meyer, *L'Union Internationale des Télécommunications*, 2.

17 Fari, *Una penisola in comunicazione*, p. 431.

were different and were linked with economics and diplomacy as it seems to have been favoured by the liberal political classes in countries viewing the international telegraph as helpful in developing the free market. On a diplomatic level, the politics of Napoleon III and his strengthening the international role of France, may well have left their mark[18].

As the following section will show, WETU adopted a very similar convention to AGTU's and after the first 1855 Paris Conference met in Turin (1857) and Bern (1858). From then on its aim was to draw closer to AGTU and merge the two into the single institution of the TU.

2.4 En route to convergence

Up to this point the history of the two telegraph unions has been told in a linear order, as if the countries belonging to AGTU did not stipulate contracts with those which had created WETU and vice versa. What really happened was very different, for during the 1850s there was a gradual process of coming closer together and overlapping. First of all, bilateral treaties were also signed between countries belonging to the two different associations: Prussia and Belgium in May 1850, Austria and Switzerland in 1852[19], Austria and Piedmont in both 1853 and 1856.

18 Milza, *Napoléon III*. On the political aim of uniting AGTU and the essentially diplomatic operation with WETU, see Carrè, "Archéologie d'une Europe des télécommunications," 76.

19 This agreement was signed to answer the need expressed by the commercial manager of San Gallo, who asked for communication between Switzerland and Austria to facilitate economic exchanges between the two countries. As the historian Verdiana Grossi records, activating this international line also answered a more wide-spread aim of the Department of the Posts and Public Works, which was to set up a telegraphic link between the North and South of Switzerland (Basel-Chiasso-Como-Milan), "a main link for commerce between the Atlantique and the Mediterranean" («liaison capitale pour le commerce entre l'Atlantique et la Méditerranée»), see Grossi, "Technologie et diplomatie," 293.

Even more significant, however, were the mixed conventions between countries from the East and West blocks, authentic "East-West connections"[20]. The first conference of this type was organised in Paris on October 1852 with Belgium, France and Prussia, the latter taking part on behalf of AGTU. A major role in this first attempt to draw the two unions together was played by Belgium, which negotiated with Prussia in the name of France, too. The convention aimed at being as similar as possible to the AGTU format stipulated two years earlier at Dresden. Two other mixed conferences were organized in Berlin (June 1855) and Brussels (June 1858). Belgium, France and Prussia took part again. This time the proactive role in merging the two unions was played by Switzerland, so the remaining part of this chapter will be taken up by an indepthing of Switzerland's position in the international context of the 1850s and the early 1860s.

As Chapter 1 records, Switzerland set up the first telegraphic network on its territory in 1852 and, right from the beginning, kept a vigilant eye on the international scene. Already in 1853, Karl F. Brunner-von Wattenwyl, Head of Swiss Telegraphs, reminded the Department of the Post and Public Works that the bilateral conventions being signed with other nations were unsatisfactory and difficult to implement, given the differences in the various national telegraph systems. For this reason Switzerland had to aim at improving international telegraphy "when it is the matter of approving conventions, we try to obtain as far as possible for international correspondence the same advantages enjoyed by the public within Switzerland"[21].

The Federal Council intervened directly one year later, in 1854. In order to replace the provisory convention agreed between Belgium, France and Prussia in the Paris Conference, it expressed a desire to organise a sort of extended European conference:

> Also, in the 1854, *we gave the idea of a European conference by inviting the administrations of the most important neighbouring states to take part in the*

20 The term is used by Lyall, *International communications*, 21.

21 «Lorsqu'il s'agit de passer des conventions on cherche à obtenir autant que possible pour la correspondance internationale les mêmes avantages que ceux dont le public jouit dans l'intérieur de la Suisse». FF, 3/ 45, 8 Octobre 1853, p. 489.

undertaking. This idea was favourably welcomed by France and Piedmont; but Austria did not deem it useful as together with Germany and Holland it is part of the Austro-German Union and already enjoys all the advantages of their common provisions.[22]

This proposal to organise a European conference to unite East and West led to several Swiss diplomatic interventions and resulted in the 1855 Berlin Conference, in which Belgium, France and Prussia took part in the name of AGTU. Its outcome was however unsatisfactory in the words of the French delegate, quoted again in the Federal Council's report to Parliament:

> His Majesty the Emperor's government delegates spent great energy on having a series of measures adopted in Berlin which would more profoundly reform the system and bring it more into line with the principles of good administration and equal distribution of costs. Yet the proposals of the French Commissioner, *with which the Belgian Representative on principle immediately agreed, encountered insurmountable resistance in the strict mandate that the Prussian Commissioner had received from the Austro-German Union, and by the as yet inadequate organization of the telegraph service in Germany.*[23]

22 «Aussi, en 1854 déjà, avons-nous fourni l'idée d'une conférence européenne, en invitant les Administrations des Etats voisins les plus importants à prendre l'initiative dans cette affaire. Cette proposition fut accueillie favorablement par la France e les Etats-Sardes; tandis que l'Autriche n'en reconnaissait pas l'opportunité, car ce pays qui, avec l'Allemagne et la Hollande, fait partie de l'Union des télégraphes austro-germanique, jouit déjà de tous les avantages de dispositions communes». FF, 1/4, 9 Janvier 1856, p. 110, [*our italics*].

23 «Les Délégués du Gouvernement de sa Majesté l'Empereur ont essayé à Berlin de faire adopter une série de mesures destinées à réformer plus profondément le système et à le mettre plus en rapport avec les principes de bonne administration et d'équitable répartition des frais. Mais les propositions des Commissaires français auxquels le Représentant de la Belgique s'était immédiatement rallié en principe, *rencontrèrent une résistance insurmontable dans le mandat restreint que le Commissaire prussien avait reçu de l'Union austro-allemande*, et dans l'organisation encore trop incomplète du service télégraphique en Allemagne». *Ibidem*, p. 111, [*our italics*].

2.5 Paris 1855: the birth of WETU

The unsatisfactory results of the Berlin convention drove the French Government to open up to other interested states and organize in December 1855 another conference with France, Belgium, Spain, Piedmont and Switzerland, and there WETU was founded.

The Federal Council furnished its delegate Brunner with precise instructions on the objectives Switzerland expected to reach, specifically in this conference and more generally at the heart of the French-speaking Union. First of all, Switzerland wanted to secure a generalized reduction of telegram tariffs, which could be obtained by revising the system of zones and more specifically enlarging them[24]. As its telegraph tariffs were considerably lower than those of other countries, Switzerland promoted reductions to line up the rest of Europe with same prices. Furthermore, a generalised reduction drop in the charges would favour international communications, which was of great interest to Switzerland.

Another concern for Switzerland was to streamline the service, given its conviction that the use of the electric telegraph could be promoted internationally by reducing drastically bureaucratic difficulties and system incompatibilities.

The third and last point on which Brunner received detailed instructions was that he was to do his utmost to ensure that convention signed at the end of this meeting was as similar as possible to AGTU's. In short, right from the beginning Switzerland lobbied for WETU, the new union to come out of the Paris Conference and comply with the German-speaking organization[25].

The three-week long Paris Conference brought important results for international telegraphy. First of all, as already mentioned, was the birth of WETU, "a *kind of telegraph union* between Belgium, Spain,

24 With a zone tariff system, an extra charge was put in each time a telegram crossed over a zone border. In this way an increase in zone size meant fewer zones and ultimately lower telegram charges.

25 On the Federal Council's instructions to its delegate at the Paris Conference see FF, 1/4, 9 Janvier 1856, p. 111–112.

France, Piedmont and Switzerland, a telegraph union which, we already know, Tuscany will soon adhere and join in, and which will later admit all the European states which will want to belong to it"[26].

Secondly, the contents of the new convention was structured so as to be almost identical to the Berlin agreement signed a few months earlier between France, Belgium and Prussia, to the point in which it was mooted that the same enumeration of the articles had to be respected[27]. What actually happened was that right from the opening debates, France, disappointed by the results of the Berlin Convention, suggested going beyond it. The desire to move closer towards the states belonging to the other union took the upper hand. It was the first victory for a mainly Swiss-driven strategy that would prove successful over the years: a policy of "small concessions" that brought the two conventions to overlap.

Another Swiss request accepted during the debate was the enlargement of the six WETU tariff zones (the first, for example, passed from 80 to 100 kms, the second from 200 to 250, and the third from 360 to 450 and so on). As was explicitly recognised by the Federal Council, "These modifications, especially substantial over large distances, are a concession by the big states to the smaller ones"[28] – a concession, in fact, that France was allowing to countries like Switzerland and Belgium. For these smaller states enlarging the zones meant that it would be less expensive to send telegrams to cities in neighbouring countries. Switzerland, for example, was interested in communicating cheaply

---

26 «une *sorte d'union télégraphique* entre la Belgique, l'Espagne, la France, la Sardaigne et la Suisse, union télégraphique à laquelle, nous le savons déjà, se ralliera bientôt la Toscane par un acte d'adhésion, et dans laquelle seront admis successivement tous les Etats de l'Europe qui voudront en faire partie». Paris International Conference Sitting, 5 December 1855, in Swiss Federal Archives (henceforth SFA), Fond E52, Archive n. 440, Band n. 2 (henceforth E52, 440/2). On the same subject see also *Légation de France en Suisse à le président de la Confederation Suisse docteur Furrer*, 18 Octobre 1855 (SFA, E52, 440/2).

27 A. Descalzi, "Creacion y desarollo de la Union Internacional de telecomunicaciones," in *Las comunicaciones entre Europa y América, 1500–1993*, ed. B. Magro, M. Llorente, O. Carvajal (Palacio de Congresos de Madrid: Actas del I Congreso Internacional de Comunicaciones, 30 de noviembre – 3 de diciembre).

28 «Ces modifications, surtout sensibles pour les grandes distances, sont une concession des grands Etats à l'égard des Etats moindres». FF, 1/4, 9 Janvier 1856, p.112.

with France and Northern Italy and with the new enlarged zones sending telegrams to French and Italian cities (particularly Genoa, Marseilles, Paris, Calais and Le Havre) would not cost as much[29].

Switzerland also managed to impose its line of thought on tariff reductions. It was in fact Brunner who explained to the delegates "the financial advantages which would come from a reduction in long distance tariffs", so managing to convince them to modify the bases of the French project and lower charges drastically[30]. WETU tariffs ended by being appreciably lower than AGTU's, and in this way Switzerland obtained a result which was a strategic high for its government.

The last issue debated at the Conference was the standardization of the telegraph apparatus for new entries. As the Federal Council had requested, a simplification of the telegraph technologies was undertaken, like the Morse telegraph being adopted for international correspondence. This was not so automatic, if we think that France had abandoned its own equipment in favour of Morse technology only in 1854, and the Swiss government found it in fact opportune to underline the fact (together indirectly with the victory of its own line): "The Morse system adopted in Switzerland will be taken up for international correspondence among all the signatory States, and therefore France is giving up its own system calling for a transcription of telegrams at border offices"[31].

The Federal Council was, of course, very satisfied with the Conference's outcome. Switzerland had obtained an all-round victory[32], showing above all it possessed a great international prestige and a notable capacity to guide both the debate and the decisions taken. Another

29 *Ibidem,* p. 113.

30 «avantages financiers qui résulteraient d'une réduction de la taxe sur les dépêches à longue distance», Paris International Conference Sitting, 10 Décembre 1855.

31 «Le système Morse, adopté en Suisse, sera appliqué à la correspondance internationale entre tous les Etats contractants, et par conséquent la France renonce à son système propre, qui nécessitait une transcription des dépêches aux stations d'échange». FF, 1/4, 9 Janvier 1856, p. 112.

32 The Federal Council held that our delegate managed to get accepted all the modifications of the project favurable to Switzerland. «Notre délégué réussit à faire accueillir *toutes* les modifications du projet qui sont favorables à la Suisse». *Ibidem,* p. 112, [*our italics*].

reason for the Swiss Government's satisfaction lay in the way it had taken part so actively in making the first moves towards a European telegraph union. It was judged as such both by Brunner, who declared his belief in the future: "I therefore hope that the agreement the commission has just accorded will one day take the name of European Convention"[33], and the Federal Council, which remembered it as "a major step towards standardization in telegraph relations throughout Europe"[34].

Figure 7: Karl F. Brunner-von Wattenwyl, Head of Swiss telegraphs in the 1850s and one of the first Swiss to have in mind an international union for regulating the services.

Source: Grossi, "Le role international de personalités suisse," 46.

33 «j'ai ensuite l'espoir que l'arrangement dont la commission vient de poser les bases porterait un jour le nom de Convention européenne». Paris International Conference Sitting, 18 Décembre 1855, handwritten.

34 «*étant* un grand pas vers l'uniformité dans les relations télégraphiques de toute l'Europe». FF, 1/4, 9 Janvier 1856, p. 112.

2.6 Turin 1857: the invite to Austria

It was the Turin Conference of May 1857 that gave Switzerland the key role as mediator between the two unions. In November 1856 Piedmont, host of the following year's WETU conference, mandated Switzerland to invite Austria to take part in the proceedings[35], i.e. gave it the delicate task of building a bridge between WETU and AGTU, on that occasion represented by Austria. The Swiss Post Department informed the Federal Council of the mandate, only to find the latter had already been making moves: "The Department of the Post and Public Works has already approached the Head of the Austrian Telegraphs to invite him to adhere to the Convention of 29 December 1855. The Austrian Administration thus finds itself informed of the wishes of the Federal Government and its opinion on the matter"[36].

Negotiations proceeded in the following months, always thanks to Swiss intermediation, and the Turin meeting was even postponed in order to understand if Austria desired to take part. Switzerland praised this delay in a telegram sent to France: "We thoroughly approve of your suggestion to put off the Turin meeting until we have some certainty about the result of the negotiations with the states of the Austro-Germanic Union. Do you not think it would be useful to approach Austria?"[37]. In spite of this readiness to meet Austria half-way, the

35 *Legation de S.M. Le Roi de Sardaigne en Suisse à Monsieur Fornerod president de la Confederation*, Berne, 30 Novembre 1856, in SFA, E52, 441.

36 «Le Département des Postes et des Travaux publics a déjà fuit des démarches auprès de la Direction I. R. des télégraphes Autrichiens pour l'engager à adhérer à la Convention du 29 Décembre 1855. L'Administration Autrichienne se trouve par le fait déjà informée du désir et de la manière de voir du Gouvernement fédéral à ce sujet». *Département des Postes et des Travaux publics de la Confédération Suisse au Conseil fédéral Suisse*, Berne, 8 Décembre 1856, in SFA, E52, 441.

37 «Nous approuvons complètement votre proposition de retarder la réunion de Turin jusqu'à ce que nous avions quelque certitude sur le résultat des négociations avec les états de l'union austro-germanique. Ne croyez-vous pas qu'il serait utile de faire des démarches auprès de l'Autriche?». *Télégramme du Département fédéral*

negotiations failed, even though Prussia appears to have been in favour of Austria being represented in Turin[38].

The Swiss Post Department had its own ideas about why the negotiations failed. "We have reason to think that the motives preventing Austria from accepting Sardinia's invitation were of an essentially political nature and go back to the strained situation existing between the two countries"[39]. In the years both preceding and following the Turin Conference there was indeed a high level of military and political tension between Austria and Piedmont, caused mainly by the latter's manœuvres to take Lombardy and Veneto out of the Austrian Empire. Major incidents included the First Italian War of Independence (1848–9) which saw the two states in opposition, and the Crimean War (1855), which Piedmont entered on the Anglo-French side, as well as the 1858 French-Piedmontese anti-Austrian agreement and the 1859 Second Italian War of Independence, with a French/Piedmontese alliance against Austria. Obviously the international alliances, tensions and strategies all had considerable influence on the development of the telegraph in Europe and in particular on the creation of a supranational body to guide the evolution of telegraphs on a continental scale.

Once the idea of inviting Austria had faded, the Turin Conference was held in May 1857 with the same countries that had participated two years earlier in Paris. WETU's main aim was, perhaps for the first time so apparent[40], to reach a common European telegraph organization and

---

des Postes au Directeur Général des Télégraphes Paris, 29 Décembre 1856, in SFA, E52, 441.

38 *Légation de S.M. Le Roi de Sardaigne en Suisse à Monsieur Fornerod president de la Confederation*, 30 Mars 1857, in SFA, E52, 441.

39 «Nous avons bien de croire que les raisons qui ont empêché l'Autriche d'accéder à l'invitation de la Sardaigne étaient d'une nature essentiellement politique et resalaient de la situation tendue qui existait entre les deux pays». *Département des Postes et des Travaux publics de la Confédération Suisse au Conseil fédéral Suisse*, 17 Août 1857, in SFA, E52, 443/2.

40 Horrenberger (*L'Union Internationale des Télécommunications*, 11) holds that the idea to merge the two unions came out for the first time at Turin, though this interpretation does not really correspond to fact. The idea of merger was already in the air, while the Turin Conference marked an important moment in which WETU members came to realize the need for a telegraphic area common to all Europe.

this was declared during the opening of the works, when "The members of the Conference agreeing on the advantages which would come with telegraphy arriving at an organization common to all the States on the continent"[41]. To do this, an agreement had to be reached with AGTU:

> The members of the Conference, with their growing awareness of how useful it would be for telegraphy to have an organization gathering together all the nations on the Continent, express the hope that efforts will be made in this direction with the members of the Austro-German association meeting in Brussels again this year.[42]

In fact, shortly before the beginning of the Conference, the French and Swiss governments agreed that one of the major aims of the meeting would be to fix another conference with the German states taking part in their own right[43]. On presenting the results of the Conference to Parliament, the Federal Council insisted in particular on standardizing the use of the telegraph on a European level as one of the objectives of WETU: "The results of this conference were not immediate, but the desire to see a complete standardization in the use of European telegraphs was again unanimous"[44]. Also in light of the desire to reach an agreement with AGTU expressed at Turin, the Federal Council was spurred on by the Federal Council to act as mediator between the two associations. Switzerland then offered itself as the geographical site between the two parties, and Brunner declared, "that he had received instructions to solicit

---

41 «Les Membres de la Conférence, convenaient des avantages qui résulteront pour la télégraphie d'arriver à une organisation commune à tous les Etats du Continent». *Legation de S.M. Le Roi de Sardaigne en Suisse à Monsieur Fornerod president de la Confederation*, 6 Juin 1857, in SFA, E52, 441.

42 «Les Membres de la Conférence, appréciant de plus en plus l'utilité qui résulterait pour la télégraphie d'arriver à une même organisation pour tous les Etats du Continent, *expriment le voeu que des efforts soient tentés à cet effet auprès des membres de l'Association austro-germanique*, qui doivent se réunir à Bruxelles dans le courant de l'année». Turin International Conference Sitting, 21 Mai 1857, handwritten, in SFA, E52, 441, [*our italics*].

43 *Lettre écrite par le Directeur Général de l'administration des lignes télégraphiques françaises adressée à Brunner*, 6 Mai 1857, in SFA, E52, 441.

44 «Les résultats de cette conférence ne furent pas immédiats, mais le désir de voir se réaliser une uniformité complète dans l'emploi des télégraphes européens se fit jour de nouveau d'une manière unanime». FF, 2/61, 30 Décember 1858, p. 687.

the Conference to choose Bern as the venue for the next meeting"[45]. The other WETU states welcomed Switzerland's candidature because it was in an ideal geographic position to function as a *trait d'union* with AGTU. This was mentioned by Brunner himself, who betrayed a certain confidence in a positive outcome of the Swiss mediation at Bern: "he was pleased about this choice above all for the positive results it could produce in perhaps facilitating the merger of the Paris Convention with the Austro-Germanic Union, which according to its tradition will be represented at Bern by one of the states neighbouring Switzerland which are all in favour of the merger"[46].

2.7 Stuttgart 1857 and Brussels 1858

AGTU met again in 1877 at Stuttgart and concluded with "a new convention which was very close to the Paris agreement of 29 December 1855"[47]. The merit for this rapprochement of the German-speaking world and WETU was Switzerland's, because in the pre-Conference months the Federal Council had suggested a text for the convention that verged on the Paris format and which would later provide the basis for the conventions of Brussels (June 1858) and Bern (September 1858)[48].

45 «qu'il avait l'ordre d'après ses instructions de prier la Conférence de fixer la ville de Berne pour le lieu de la prochaine réunion». Turin International Conference Sitting, 21 Mai 1857, handwritten, in SFA, E52, 441.

46 «il se réjouit de ce choix surtout pour les heureux résultats qu'il peut produire en facilitant peut-être la fusion de la Convention de Paris à l'Union Austro-allemande, qui selon l'usage qu'elle a adopté se fera représenter à Berne par l'un des Etats limitrophes à la Suisse qui sont tous favorables à la fusion». *Ibidem.*

47 «une nouvelle convention qui se rapprochait essentiellement de celle dite de Paris, du 29 Décembre 1855». FF, 2/61, 30 Décembre 1858, p. 688. This Conference is well known for introducing the custom of drawing up two types of document, concerning the conventions and the regulations with services instructions, see Codding, *The International Telecommunication Union*, 15.

48 FF, 2/61, 30 Décembre 1858, p. 688.

The June 1858 conference was "mixed" with Belgium, France and Prussia taking part, the latter in name of AGTU. Following the Swiss line they "concluded a convention which contained the main conditions of the Stuttgart Convention"[49]. According to the Federal Council, it was in fact with the Brussels meeting that France started to share Switzerland's point of view and began to promote a European telegraph union:

> France, seconding our point of view, sent us a draft convention based on the Paris one but which contained the same provisions as Stuttgart, which we ourselves proposed and which were adopted in Brussels.[50]

2.8 Bern 1858: an attempt to clone Stuttgart

In the months running up to the Bern Conference the Federal Council made a notable diplomatic effort to invite AGTU members and in particular delegates from Austria, Württemberg and Baden, with whom Switzerland was on good terms[51]. Belgium, France,

49 «avaient conclu une convention qui contenait les principales conditions de la convention de Stuttgart» *Ibidem.*

50 «La France, *entrant d'ailleurs tout-à-fait dans nos vues*, nous adressait un projet de convention qui avait pour base la convention de Paris, mais qui renfermait ces mêmes dispositions de Stuttgart, *que nous avions nous-mêmes proposées* et qui avaient été adoptées à Bruxelles». *Ibidem.*

51 *Département des Postes et des Travaux publics de la Confédération Suisse au Conseil fédéral Suisse*, Berne, 17 Août 1857, in SFA, E52, 443/2. See also "On 17 August 1857 The Federal Council refrained from proposing for the Austro-Germanic Union via the intermediation of the governments of Austria, Wuttenberg and Baden to take part in the Conference under way in Bern in order to achieve a standardized orgaization of telegraphs" («le Conseil fédéral s'est empêché, le 17 Aout 1857, de proposer à l'Union austro-germanique, par l'intermédiaire de le gouvernement d'Autriche, de Wurtemberg et de Bade, de prendre part à la conférence actuellement réunie à Berne dans le but d'obtenir une organisation uniforme des télégraphes»). Bern International Conference Sitting, 24 Août 1858, handwritten, in SFA, E52, 443/1.

Piedmont, Portugal, Holland, Württemberg and Baden were all present[52].

As in Turin, the main aim was to bring the two conventions as close as possible. Switzerland mediated as always and stressed that both groups had to follow the path already traced out. The premises were good and showed signs of leading to the desired result. The confidence of both delegates and the Swiss government came in the first place from a justified optimism: "the spirit that reigned in the Turin conferences leaves [...] no doubt about the fact that the contracting States would be ready to accept certain modifications which would seem essential for the merger which is objective of all"[53]. Secondly, as Wilhelm Matthias Naeff, Head of Swiss Post and Public Works, declared during the opening of the conference, Swiss diplomatic activity in the previous months had been well accepted by both groups, particularly AGTU[54].

During the debate, the aim was to make the convention as similar as possible to those signed by the AGTU countries. Belgium suggested, for example, starting from the Brussels text rather than the Paris one because the former was "in line with the Paris project on the general principles, but also has the advantage of bringing together in a more complete way the Western and Austro-Germanic groups"[55]. In order to reach a complete integration between the various positions, France defended the project it had drawn up for the Bern Conference, but suggested examining carefully the Brussels text because it was based on the Stuttgart model.

52 Holland, Wurtenberg and Baden, all AGTU members, took part in the Bern Conference in a personal capacity.

53 «l'esprit qui a régné dans les conférences de Turin ne laisse [...] aucun doute sur ce fait que les Etats contractants seraient disposés à accepter certaines modifications qui paraîtraient nécessaires à la fusion générale qui fait l'objet de leurs vœux unanimes». *Département des Postes et des Travaux publics de la Confédération Suisse au Conseil fédéral Suisse*, 17 Août 1857, in SFA, E52, 443/2.

54 Bern International Conference Sitting, 24 Août 1858, handwritten, in SFA, E52, 443/1.

55 «se trouve d'accord, dans ses principes généraux avec le projet de Paris, mais présente sur ce dernier l'avantage de concilier entre d'une manière plus complète les deux groupes de l'Union occidentale et de l'Union austro-germanique». *Ibidem*.

In a message to the Swiss Parliament, the Federal Council expressed a certain satisfaction at the outcome of the Bern Conference, and underscored the role played by Switzerland in reaching these aims:

> The most important of these general agreements was the marked tendency to converge in a European telegraph unity, a move that the Swiss telegraph administration was one of the first to propose during the Paris telegraph conference, which it has promoted since then, on which it has worked with all its might, which is finally beginning to come to life and of which it can, justifiably, be proud.[56]

According to the Federal Council again, the Bern Conference had reached the main aim it had been organized for, with the standardization of the telegraph regulations for Europe. The principal adjustments to the Paris Convention had made:

> *the Bern one so similar to those of Stuttgart and Brussels, which can be considered identical in their application*, and a single rule can be applied to all telegraph correspondence of the countries represented by the three conventions, countries which will *without doubt soon be joined by most of the other European countries.*[57]

The 1858 Brussels and Bern treaties, implemented at the same time, "created for Western Europe a kind of uniform regime"[58]. In the

---

56 «Le plus important de ces traits généraux a été la tendance marquée à se rapprocher d'une unité télégraphique européenne, que *l'Administration des télégraphes suisses a été une des premières à mettre en avant dans les conférences télégraphiques de Paris*, qu'elle a dès lors provoquée, à laquelle elle a travaillé de tout son pouvoir et qui vient enfin de recevoir un commencement d'exécution et de vie dont *elle peut, à bon droit, être fière*». FF, 2/62, 30 Décembre 1858, p. 703, [*our italics*].

57 «*celle de Berne si semblable à celles de Stuttgart et de Bruxelles, que dans leur application du moins elles peuvent être considérées comme identiques*, et qu'une seule règle pourra être appliquée à toutes les correspondances télégraphiques des pays représentés dans les trois conventions en question, pays auxquels viendront se joindre *sans aucun doute très-prochainement la plus grande partie des autres Etats européens»*. FF, 2/61, 30 Décembre 1858, pp. 690–91, [*our italics*].

58 «constituaient pour l'occident de l'Europe une sorte de régime uniforme», see E. Saveney, "La télégraphie internationale. Les anciens traités et la conférence de Paris," *Revue des deux mondes*, 15 Septembre 1872, 365.

following years, with the essential backing of Switzerland, this substantial uniformity of direction led to an international union, and it is by no coincidence that its first convention mirrored almost to perfection the 1858 Bern Convention[59].

2.9 Friedrichshafen 1858: Switzerland holding the balance

During the Bern Conference Switzerland put pressure on Baden and Württemberg to make them act as intermediaries and invite Austria to the next conference which was to take place at Friedrichshafen, a small town on Lake Constance. The conference, which opened on 21 October 1858, six weeks after the ratification of the Bern Conference, was officially organized by Switzerland to review the conventions in force between Switzerland and AGTU, though it turned out to be another step forward in bringing the two unions together.

Switzerland, Austria, Baden and Württemberg all took part, and on the Swiss suggestion, the freshly ratified Bern Convention was taken as starting point, even though the new convention would be naturally drawn up in German[60].

59 Lyall, *International Communications*, 24.

60 "From the beginning and on the Swiss delegates' suggestion, the conférence decided to take the Bern Connvention as the basics for the future one, which would any way always drawn up in German" («Dès l'entrée et sur la proposition des délégués suisses, la conférence décida de prendre la convention de Berne pour base de la future convention, laquelle serait toutefois rédigée en langue allemande»). FF, 2/61, 30 Décembre 1958, p. 694. Switzerland played an interpreting role between the two unions. Given its level of plurilinguism, it often promoted the translation and study of the conventions from French to German, thus familiarising the various countries with the precise terminology.

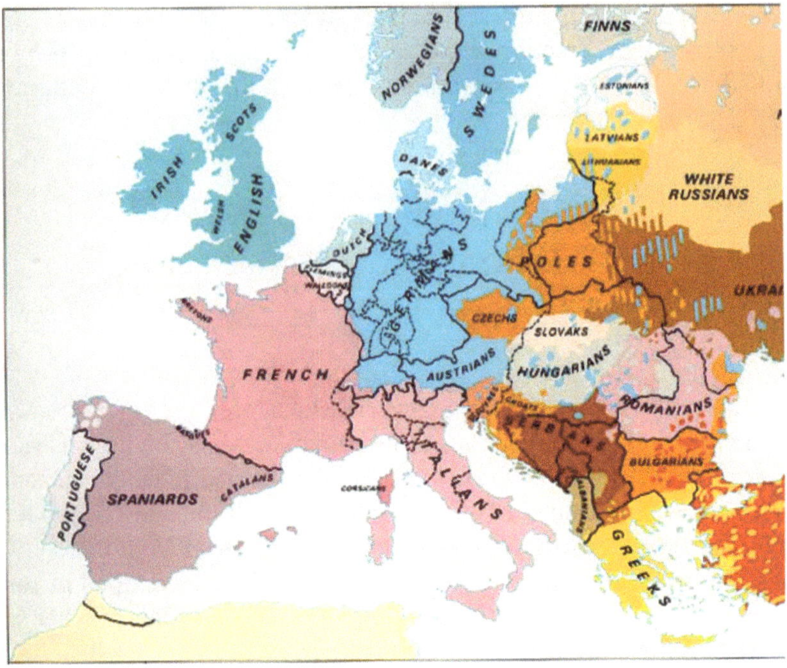

Figure 8: Switzerland also as a language mediator between the two unions.

Source: The Times "Atlas of World History", *1978, 214: Languages, peoples and political divisions of Europe [1800 to 1914], <http://cafehome.tripod.com/serbdom-eng.htm>.*

According to the Federal Council, a series of noteworthy results were reached. Firstly the countries representing AGTU were in favour of standardizing most of the main contents of the Bern Convention and it was decided to implement both conventions at the same time to avoid administrative and managerial problems: "Implementation of all the conventions concluded the same day and consequently the unitary principle underlying them; consequently avoidance of any transitory period, so troublesome in administrative matters"[61]. Secondly, Switzerland

61 «Mise en vigueur le même jour de toutes les conventions conclues et par conséquent du principe unitaire qui les domine; par conséquent, aussi évitation de tout état transitoire, toujours si fâcheux en matière administrative». *Ibidem,* p. 700.

managed to promote the standardization and rationalization of the telegraph systems both among AGTU countries and Europe as a whole: “Absolute unity in the way of applying telegram tariffs. [...] Uniform and rational system of classification of the border stations for applying the tariffs in four classes [...] Application of a single service regulation for the telegraph correspondence of the contracting States, which will shortly embrace the whole of Europe”[62]. A third objective that Switzerland reached was a considerable tariff reduction between itself and AGTU member states.

Where Switzerland was less successful was in its attempt to enlarge tariff zones as it had done with France within WETU. It was in Austria’s interests, in particular, to keep its own tariff system with smaller zones because its territory was greater and more zones meant obtaining greater revenue from international telegrams. Though with some scepticism, the Federal Council had evidently trusted it could convince AGTU to accept enlarged zones. Failing to do so led the government to question itself about what appeared a defeat both for Swiss interests and especially for achieving a full telegraph unity at European level:

> It will be noted besides that the nine states forming the Austro-Germanic Telegraph Union were most unlikely to be disposed to give up in favour of Switzerland their zone system to adopt the Bern one, whereas France and Belgium had not asked for it. It would have been necessary for Switzerland to adopt the Austro-Germanic zone system which was not necessary, apart from the point of view of equality, and would have been a step backwards on the road to European telegraphic unity that we have been pursuing for many years.[63]

62 «Unité absolue dans le mode d’application des taxes aux dépêches télégraphiques. [...] Système uniforme et rationnel du classement des points frontières pour l’application des taxes en quatre groupes [...] Application d’une instruction de service unique pour les correspondances télégraphiques des Etats contractants, qui embrasseront sous peu l’Europe entier». *Ibidem.*

63 «On remarquera en outre qu›il était peu probable que les neuf Etats qui composent l›Union télégraphique austro-germanique eussent été disposés à renoncer, en faveur de la Suisse, à leur système de zone pour adopter celui de Berne, tandis que la France et la Belgique ne l›avaient point exigé. Il aurait donc fallu que la

For correspondence with German-speaking states, Switzerland lined up with the zone system adopted by AGTU, though it was only an apparent modification and, more importantly, a temporary one:

> Besides this concession that we make in reducing the charges for a certain number of Swiss offices in their relations with the Austro-German Union, it is evidently an additional benefit for telegraph correspondence and earnings for the Swiss public, in line with the liberal principles that have always guided the Federal Authorities in the management of telegraphs.[64]

In other words, Switzerland forewent eliminating some tariff zones, but conceded to Austria lower tariffs for telegrams sent from some designated Swiss offices. In this way it experimented transmitting telegrams to Austria with the same tariffs as in a no-zone regime. The aim was to show that over time the introduction of criteria seeking to lower tariffs and therefore wipe out zones would be advantageous for Austria, too. In conclusion, Switzerland only apparently gave way to Austria's requests while it was really placing on the table instruments which would allow it to argue its own case in the near future.

The National Council Commission mandated to study the Bern and Friedrichshafen telegraph conventions in January 1859 drew up a report in which, in more than any other document analysed up to now, there emerges the major role the country played in stitching together the two European telegraph unions. First of all, the commission too recognised that the two conventions marked a significant step towards European telegraph unification, but that other steps were needed: "It is so true that even the conventions submitted today for ratification must only be considered provisory adjustments which will be replaced

Suisse adoptât le système de zones austro-germanique, ce qui, abstraction faite de ce qu›au point de vue de l›équité la chose n›est point nécessaire, eût été un pas en arrière dans celte voie de l›unité télégraphique européenne que nous poursuivons depuis plusieurs années». *Ibidem,* p. 695.

64 «D'ailleurs cette concession que nous faisons en diminuant les taxes pour un certain nombre de bureaux suisses dans leurs rapports avec l'Union austro germanique est évidemment une facilité de plus donnée aux correspondances télégraphiques et un gain pour le public suisse, conforme aux principes libéraux qui ont toujours guidé les Autorités fédérales dans l'organisation de l'administration des télégraphes». *Ibidem.*

one day by an international treaty among all the European states"[65]. The main aim of the two conferences, to obtain two almost identical conventions had been reached, thanks above all to the Swiss delegates who had worthily completed the delicate mission they had been assigned [66]. All things considered, with the ratification of the Bern and Friedrichshafen conventions, Switzerland, strengthened by its neutrality that appeared as a sure sign of impartiality also in the telegraph sector, saw the first meaningful fruits of the efforts it had made back in the 1850s to merge the two telegraph unions, which are rechristened significantly "the two great large electric currents dividing Europe"[67].

> The Bern conference took a giant step towards this desirable goal. The task was not at all easy. Indeed, it *meant bringing together the two large electric currents that divide Europe.* On the one hand France and the western states; on the other, Austria and the member states of the Austro-German union. Each of these two great powers wanted their own system to prevail. *Switzerland, a neutral country, was tasked with finding an agreement and, if possible, unification.*[68]

Following the ratification of the treaty between Switzerland and AGTU (represented by Austria, Baden and Württemberg), there no longer existed any reason for the whole German-speaking union not to ratify the 1858 Bern Convention. In the name of AGTU Prussia had already signed a convention with France and Belgium at Brussels in the same year, a convention based on the Stuttgart one AGTU had stipulated in 1857. Also in the name of AGTU Austria signed the 1858

65 «Cela est si vrai que même les conventions soumises aujourd'hui à notre ratification ne doivent être considérées que comme des arrangements provisoires, qui feront place un jour à un traité international entre tous les Etats européens». FF, 1/6, 5 Février 1859, p. 110.

66 «dignement rempli la mission délicate dont ils étaient chargés». *Ibidem.*

67 «les deux grands courants électriques qui se partagent l'Europe». *Ibidem.*

68 «La conférence de Berne a fait faire un grand pas vers ce but désirable. La tâche à remplir n'était point facile. *Il s'agissait en effet de rapprocher et de réunir les deux grands courants électriques qui se partagent l'Europe.* D'un côté la France et les Etats occidentaux, de l'autre l'Autriche et les Etats qui constituent l'Union austro-germanique. Chacune de ces deux grandes influences voulait faire prédominer son système particulier. *A la Suisse, pays neutre, il incombait à amener l'entente et, si possible, l'union». Ibidem,* [*our italics*].

Friedrichshafen Convention, which was practically identical to the 1858 Bern version signed by WETU countries. AGTU had substantially taken over the most significant informative principles of the conventions of the French-speaking area and so joined the 1858 Bern Convention on 1 March 1859[69].

2.10 Bregenz 1863: a European Union in view

Five years on from the Friedrichshafen Conference, Switzerland signed another telegraph treaty with AGTU, this time with Austria, Württemberg, Baden and Bavaria. There were no great changes or novelties in subject matter, except a Swiss attempt to introduce a single tariff for all telegrams to and from AGTU states. It was not successful, but at least Switzerland obtained a reduction in the number of AGTU zones, which went down from ten to four. It had little effect on Swiss interests since its citizens mainly exchanged telegrams only with the first and second zones, where tariffs remained substantially the same[70].

As far as progress towards the creation of a single European telegraph association was concerned, the Bregenz Conference was in the Federal Council's eyes a step ahead. Louis Curchod, the new head of Swiss Telegraphs after Charles Brunner took over the management of the Austrian network in 1857, sent the Swiss Post Department a memorandum holding that many of the AGTU states were in favour of forming a European telegraphic union and therefore the moment was ripe for accelerating operations:

> For myself I insisted on the urgency there would be in replacing as quickly as possible the various existing international treaties with a single European agreement.

69 Durand Barthez, *Union Internationale des Télécommunications*, 32.

70 *Rapport de Louis Curchod au Conseil Fédéral sur la Conférence de Bregenz*, Berne, 7 Novembre 1863, in SFA, E52, 444.

> This urgency was recognised by all the delegates and a declaration on the subject was put in the protocol, in virtue of which the Federal Council will be able to rely on the wishes of Austria, Bavaria, Baden and Wuttenberg to bring about the meeting of the general conferences in Paris, just as foreseen in the treaties of Bern and Bruxelles.[71]

Similar words would be used a few weeks later by the Federal Council, which in a message to Parliament of November 1863, expressed its own optimism about the convocation of a future European telegraph conference:

> On the contrary, we pin our hopes on the fact that with this first step [...], a European treaty which supersedes all the other international treaties will soon be signed in the general conferences in which we will invest all our energy.[72]

In fact in December 1863 Switzerland signed an agreement with France to lower the tariffs for the telegraphs between the two countries and on the occasion:

> The French administration announced that it intended to make constant use of this right and that it would convene at the same time in Paris the general conferences foreseen in the said treaty so as to replace the last with an international treaty that would bring together the highest number of European states as possible.[73]

---

71 «j'insistai alors de mon côté sur l'urgence qu'il y aurait à remplacer dans le plus bref délai possible les divers traités internationaux en vigueur par un seul traité européen. Cette urgence fut reconnu par tous les délégués et une déclaration y relative fut inscrite au protocole, en vertu de laquelle le Conseil fédéral pourra s'appuyer des vœux de l'Autriche, de la Bavière, de Baden et du Wurtemberg, pour provoquer s'il y a lieu la réunion de conférences générales à Paris, ainsi qu'elles sont prévues par les traites de Berne et de Bruxelles». *Ibidem.*

72 «nous avons au contraire tout espoir, ce premier pas [...], qu'un traité européen, supprimant tous les autres traités internationaux, ne tardera pas à être conclu dans des conférences générales auxquelles nous pousserons de tout notre pouvoir». FF, 3/55, 19 Décembre 1863.

73 «L'Administration française annonça qu'elle entendait faire incessamment usage de ce droit et qu'elle convoquerait en même temps à Paris les conférences générales prévues par le dit traité, afin de remplacer ce dernier par un traité international réunissant le plus grand nombre possible des Etats de l'Europe». *Ibidem*, p. 934.

The Swiss government naturally learnt the news with pleasure because it would achieve an objective that it had been following for a decade:

> We can only be satisfied about the possibility of the general conferences which, we do not doubt, *will complete the structure of which we laid the foundations in the various telegraph conventions* now submitted for the approval of the Higher Federal Assembly and which could also, as regards the question of the border offices *contribute to bringing the French Administration closer to our ideas.*[74]

2.11 Switzerland's solutions

Switzerland's role in converging AGTU and WETU members into the international telegraph union was crucial. The association was indeed born thanks to mainly the shrewd strategy that Swiss government employed, which turned out to be outstanding for three reasons.

First of all Switzerland made the most of its status of a neutral country with a reputation for international mediation and often insisted that a structure as complicated as the creation of an international union could only come about by means of «reciprocal concessions»[75].

A second strategy Switzerland deployed to resolve the differences between the two unions was an inch-by-inch policy. It always sought to intervene so that the conventions agreed upon in the various AGTU and WETU conferences were more and more similar so that they ended

74 «Quant à nous, nous ne pouvons voir qu'avec satisfaction l'éventualité de conférences générales qui, nous n'en doutons *pas, viendront couronner l'édifice dont nous venons de poser les fondements par les diverses conventions télégraphiques* soumises maintenant à l'approbation de la haute Assemblée fédérale, et qui pourraient bien aussi, eu ce qui concerne la question des bureaux frontière, *contribuer à convertir l'Administration française à nos idées». Ibidem*, [*our italics*].

75 This term was used by the Federal Council Commission which evaluated the Bern and Friedrichschafen conventions: «Il en résulta que la convention de Berne put à son tour servir de base à la nouvelle conférence qui eut lieu à Friedrichshafen, et qu'ainsi, *au moyen de concessions réciproques*, on arriva à formuler deux conventions presque identiques». FF, 1/6, 5 Février 1859, p. 110.

by pratically overlapping. A strategy like this, inevitably linked to the tactic of reciprocal concessions, needed time to be realized and in fact the process uniforming the conventions of the two unions took over ten years. The "slowly-slowly" process seemed the only way to bring to convergence two groups of countries which were already far from homogeneous internally, and also very distant from one another in political, economic, technical and social terms. Upturning the conventions at every meeting and identifying preferences and getting adverse parties to agree would have been a recipe for diplomatic disaster. The Swiss *modus operandi* sought however to move and manoeuvre with small but meaningful steps, learning lessons from the past, alerting other nations to the need to set up an international body overseeing European telegraph traffic.

The third strategy Switzerland adopted belonged more to the political sphere than to technical know-how in telecommunications. In those years of unrest and war in Europe, Switzerland was one of the few countries which could boast of discreet relations with the Austrian Empire. With the concessions made to Hungary after the 1848 uprisings, Austria had lost much of its German character and had become a multi-ethnic state in a moment in which senses of national identity were becoming stronger and stronger. The Austrian Empire was an obstacle for the forces aspiring to the political unification of firstly the Italian and German peoples and later the Slavs. The relations of Austria with Prussia and Italy (and also with France because of its expansionist policies) were strained, so that Switzerland, having no conflicting interests, could safely keep to its general neutral policy. It may also have helped that Switzerland had some political differences with Prussia, which went to strengthen the axis with Austria. In the dying days of 1856, the Prussian King Frederick William IV had threatened to intervene in Switzerland over the Neuchâtel Affair, which was avoided thanks to the diplomacy of Johann Konrad Kern, a Swiss member of parliament who was later to play an active role in the telegraph sector too[76].

---

76 See C. Altermatt, "On Special Mission: Switzerland and its Diplomatic System," in *The Diplomats' World. A Cultural History of Diplomacy, 1815–1914*, ed. M. Mösslang and T. Riotte (Oxford: Oxford University Press, 2008), 326–327.

Finally, what helped cement the relation between Austria and Switzerland was an extemporary event, which turned out to be very important in a universe populated by technocrats famous also for their political nous. Charles Brunner, head of Swiss telegraphs and initiator of the Swiss pro-Europe policy, took up the post of Head of Austrian Telegraphs in 1857. His closeness to the Federal Council and friendship with his successor Curchod must have been decisive in helping bring the two countries close[77].

The other powers also realized that creating a European union without Austria would have been impossible, both because of the size of its territories and because it was strategically important as a doorway to the Orient. Switzerland was the main go-between for WETU countries and Austria and also between the latter and a part of the AGTU countries. The special role began to be very visible when Piedmont requested the Federal Council to persuade Vienna to take part in the 1857 Turin Conference. Switzerland's privileged relation with Austria then emerged in all its relevance in the months leading up to the first TU conference in Paris, because it balanced out an alliance which had been formed in the early 1850s between Belgium, France and Prussia, in the two "mixed conventions" of 1852 and 1855. The other point of contact between WETU and AGTU had been made thanks to Switzerland via the 1858 Friedrichshafen Convention. Bringing Austria close to the positions of the Bern Convention, Switzerland realized a double objective: firstly it laid the basis for the second biggest AGTU country to be included in the merger and, secondly, bringing in Austria and later on Russia meant opening a door to the Orient and therefore the colonies in Asia.

2.12 Diplomacy and imbroglio in the 1850s and 1860s

According to the historian Durand Barthez, the decade leading up to the birth of the TU in 1865 was marked by a kind of "diplomatic imbroglio"

77 Grossi, 'Technologie et diplomatie'.

with numerous bilateral agreements between the various countries and similar, but still separate, conventions signed by the two unions compounded by a more and more pressing need to standardize international telegraph norms[78]. The Telegraph Union can also be seen as the best solution for all these problems or as has been held a kind of "natural outlet for all the conventions and treaties that had preceded it"[79]. In fact both administratively and technically, the TU's new normative and technical structure was to be a simplification or better still a rationalisation of all previous treaties. Probably, such a process of rationalisation was favoured by the fact that the conference delegates were not simply delegates, but also specialised engineers and technicians. The technical delegates, often heads of state telegraph administrations, concentrated on resolving the issues most closely associated with the services they directed and nurtured little interest in the political questions outside their spheres of competence. The technicians saw no problem in overriding possible national conflicts in order to make the telegraph service more efficient as this was their professional task as state employees. Furthermore, the technicians easily understood one another since telegraph services were technically similar everywhere. They also had the same mind cast, as most of them were engineers and managers more used than diplomats to concentrating on goals and discarding distracting detail. This is why the technical elite of top telegraph managers can be reputed to have played an important role in the rapid resolution of a diplomatic imbroglio that had been snowballing over the years.

78 Durand Barthez, *Union Internationale des Télécommunications*, 32.

79 «sbocco naturale di tutte le convenzioni e di tutti i trattati che l'avevano preceduta», see Fari, *Una penisola in comunicazione*, 430.

Chapter 3
The birth of the Telegraph Union: the 1865 Paris Conference

Introduction

Apart from the American Civil War taking place between 1861 and 1865, there were in the early sixties at least three decisive events for the political and economic destiny of Europe. In the first place, Italy was unified in 1861, so ending the foreign occupations that for years had divided up its territory and at the same time, creating a single new national interlocutor[1]. Secondly Otto von Bismarck was elected Prussian Chancellor in 1862 and acted at once to oppose Austria and remove its influence from German territories, so becoming the main protagonist of German unification of 1871[2]. Thirdly and importantly, the early 1860s were the period of Napoleon III's greatest activity in foreign affairs.

This third event, for many reasons the most important from the viewpoint of creating an international body to regulate continental communications, merits being indepthed especially for two reasons. First of all, Napoleon III undertook action supporting European national movements with the aim of questioning the borders established by the Congress of Vienna and restoring France's leading role on the Continent. In particular, in 1859 France allied with Piedmont in the victorious war against Austria for the possession of Lombardy and laid the basis for

1 D. Mack Smith, *Modern Italy: A Political History* (Ann Arbor: University of Michigan Press, 1997).

2 A. J. Percival Taylor, *Bismarck: The Man and the Statesman* (London: Hamish Hamilton, 1955).

the creation of the Italian confederate state[3]. Less fortunate his policy of backing Prussian expansionism – directed at obtaining in exchange freedom of action in Belgium, Luxembourg and the Palatinate – and even less so the Mexican expedition which should have led to the creation of an empire under Maximilian of Habsburg[4]. Secondly, France began a very aggressive colonial policy in Africa, Indochina and China which, sanctioned a couple of decades later at the 1885 Congress which made it one of the main actors in the globalization of the European economies taking place in the second half of the nineteenth century[5]. Infrastructure planning also revealed the supranational scope of French intentions, for between 1869 and 1869 Ferdinand de Lesseps was working on the project of the Suez Canal[6].

The call for the international telegraph conference in Paris belonged to a European context in which some nations had found a new political stability while others were still searching for it, France was aiming at winning back a role as guide and in many ways beacon to the development of modernity[7]. Significantly, it almost seems, the first International Telegraph Conference took place in Paris.

3.1 Austria's presence

The bilateral treatises between the various states and the rapprochement of the two telegraph unions through the 1850s and early 1860s resulted,

3 W. Deutsch, *Habsburgs Ruckzug aus Italien: die Verhandlungen von Villafranca und Zurich 1859* (Wien und Leipzig: Adolf Luser, 1940).

4 K. Ibsen, *Maximilian, Mexico, and the Invention of Empire* (Nashville: Vanderbilt University Press, 2010).

5 M. E. Chamberlain, *The Scramble for Africa* (London: Longman, 1974).

6 D. McCullough, *The Path Between the Seas: The Creation of the Panama Canal, 1870–1914* (New York: Simon and Schuster, 1977).

7 Note that in 1860, Napoleon III's regime was already named the *Empire libéral* for the attention it paid to the instances of the new middle-class on both national and international levels.

as we have seen, in the creation of the Telegraph Union. For the reasons mentioned above France offered to organize the first international conference to bring it into being. The aim was to gather together in Paris the representatives of the European countries which had agreed on a government-run, public monopoly telegraph service. Switzerland yet again had been skilfully mediating in the years running up to the conference, and in manoeuvring ceaselessly to have Austria present at the meeting, too.

On 12 July 1864 the French Minister for Home Affairs invited the Federal Council to take part in the international conference in order to "manage if possible to establish one single regulation for telegraphic correspondence in all the European states"[8]. Invitations also went out to all the nations which had taken part in 1858 Brussels and Bern Conferences. In other words, for the French, the countries which were to partake in the birth of the Telegraph Union were Belgium, France, Prussia (which had also represented AGTU in Brussels), Switzerland, Italy, Portugal, Holland, Württemberg and Baden.

Four days later, on 16 July 1864, Naeff, the chief of the Swiss Post Department, presented the Federal Council with the French invitation, observing that to bring into being a real European telegraph unity, not only[9] the delegates present at Bern and Brussels had to be included, but also the other European nations up to then not participants. Though the reference to Austria was not exactly explicit, it was certainly clear enough.

Realising that France wanted to keep Austria out of the future union, Switzerland undertook an operation of mediation/pressure to bring it into the new institution. Between late 1864 and early 1865 its strategy was developed on two different levels. In their

8 «arriver s'il est possible à établir une seule et même règle pour la correspondance télégraphique du tous les Etats européens». *Ministère de l'intérieur français à le directeur central des télégraphes suisses Curchod*, 12 Juillet 1864, in SFA, E52, 502.

9 «non seulement». *Le Département des Postes au Conseil Fédérale*, 16 July 1864, in SFA, E52, 502.

considerable correspondence, the heads of the Swiss and Austrian telegraphs (Curchod and Brunner[10]) converged on several points and established a common strategy. The Federal Council also made its weight felt and intervened politically in favour of Austria. Yet again Switzerland was stepping into the international limelight by moving on two levels: technically (with Curchod, his fame and international ties with heads from other countries) and politically (with the country deploying its proverbial and internationally recognised capacity for mediation).

After being invited, Curchod wrote to Brunner anticipating that he too would soon be receiving an invitation from the French. It is not clear if Curchod was trying to sound Brunner out or if he simply was unaware of the French manoeuvres, but a month later Brunner denied having received any invitation and added that, "if you aspire to a European system, invite all the states in the Union so that they can at once form a committee"[11].

At the end of September, Curchod answered and guaranteed that Switzerland had already made moves and would make others to obtain an invitation for Austria. In particular, Curchod let Brunner know he had already taken it upon himself to pressure the Head of French Telegraphs and the Federal Council was ready to send an official request for invitations to be sent to all the nations which had signed the 1858 Friedrichshafen Treaty, putting forward that the convention signed on Lake Constance had the same value as the Bern and Brussels ones:

> we have decided to impose again on the French Administration (what I had already had occasion to do verbally) the fact of the existence of the Friedrichshafen Convention in respect to Bregenz, uniting beside the Bern and

10 Friends, Curchod and Brunner held each other in high esteem. Brunner had been both Curchod's mentor and predecessor in the early stages of telegraphy in Switzerland.

11 «si vous aspirez à un système Européen, invitez tous les états de l'Union, pour que ceux-ci puissent immédiatement se constituer en comité». *Lettre écrite par Brunner adressée à Curchod*, 11 Août 1864, *underlining in the original document*, in SFA, E52, 502.

> Brussels groups a third group of European states which will have to be according to us convoked in Paris if we wanted to have some assurance of achieving the aim that the French Administration proposes […]. We have said then that we will be ready, were it to seem necessary to procure an official request from the Federal Council sent to the French Government aiming at obtaining that the signatory states at the Friedrichshafen (Bregenz) Convention to also be convened at the Paris conferences.[12]

The 1858 Friedrichshafen Convention had been signed by four nations: Switzerland, Baden, Württemberg and Austria. Since the first three had already been invited to Paris, Switzerland decided not to name the *omissis* explicitly but to intervene directly.

On the same day in which Curchod wrote the above letter fully supporting Brunner, he also sent a message to the Head of French Telegraphs de Vougny making the Swiss position clear. He held that besides including the states present at Brussels and Bern, those that had signed Friedrichshafen and Bregenz had to be added for a series of reasons. First of all, the countries signing the latter two conventions, and the main reference was naturally to Austria, were all in an important geopolitical position for the proper functioning of the TU[13]. The Austrian Empire bordered on nations that were joining the new European network (some German states plus Prussia,

12 «nous sommes décidé à imposer de nouveau à l'Administration française (ce qui j'avais déjà eu l'occasion de faire verbalement) le fait de l'existence de la convention de Friedrichshafen, risp. de Bregenz, réunissant à côté des groupes de Berne et de Bruxelles un 3e groupe d'Etats européens, qui devrait selon nous être également convoqués à Paris si l'on voulait avoir quelque assurance d'atteindre le but que l'Admin. Française se propose […]. Nous avons dit enfin que nous serons prêts, si le paraissait nécessaire, à provoquer une demande officielle du Conseil féd. au Gouvernement français tendant à obtenir que les Etats signataires de la convention de Friedrichshafen (Bregenz) soient aussi convoqués aux conférences de Paris». *Lettre «confidentiel» écrite par Curchod adressée à Brunner*, 26 Septembre 1864, in SFA, E52, 502.

13 Geopolitics naturally plays a key role in the social history of telecommunications. For a network analysis of European telecommunications in the nineteenth and twentieth centuries see R. Wenzlhuemer, *Connecting the Nineteenth-Century World. The Telegraph and Globalization* (Cambridge UK: Cambridge University Press, 2013), especially chapter 6.

Switzerland and Italy) and leading to Eastern Europe. For this reason it was of fundamental strategic importance, more so actually than Prussia[14]. Then, in the previous decade and specifically at the Friedrichshafen and Bregenz conferences, Baden, Württemberg and Austria had made a considerable effort to meet WETU halfway and seemed ready to make other concessions, including the introduction of a single tariff : "With the Swiss proposal for the adoption of the principle of a single tariff [...] the delegates from Austria, Bavaria and Württemberg will send to the Protocol that they consider the principle of the single tariff a very favourable objective"[15]. Finally Curchod mentioned in a fairly cryptic way that Switzerland was interested in "not compromising its independence by behaving in an indecisive vague way at Bregenz"[16]. With this formulation Curchod wanted to recall at least three issues – Switzerland would refuse to sign any conventions that would force it to relinquish national sovereignty over telegraphs; the absence of a nation with which it had signed a previous treaty would curb possibility of manoeuvre; it would defend what had been decided at Bregenz, partly in order not to lose its freedom and autonomy.

In the conclusion of this long letter, Curchod addressed a series of rhetorical questions to his colleague to get him to understand that without Austria (and with it an important part of AGTU), the whole new telegraph union would be undermined:

14 *Lettre écrite par Curchod adressée à de Vougny*, 26 Septembre 1864, in SFA, E52, 502.

15 «La Suisse ayant proposé l'adoption du principe de la taxe unique [...] les délégués de l'Autriche, de la Bavière et du Wurtemberg consigneront au protocole, qu'ils considéraient ce principe de la taxe unique comme un but qu'il était très désirable d'atteindre». *Ibidem*.

16 «de ne pas compromettre son indépendance eu s'engageront à Bregenz d'une manière indéterminée». *Ibidem*.

Figure 9: The Austrian telegraph network in Europe, 1860: the international telegraph network's doorway to the east. Map taken from *The Telegraphs of Europe* published by the Electric & International Telegraph Company in 1860 and compiled by Francis Young.

Source: <*http://distantwriting.co.uk/telegraphmap.aspx*>.

> Would it not be fitting on the contrary that all the Union states were represented at Paris […]? Would it not be fitting, above all in the interests of the principles now prevailing in Western Europe, not to neglect the support of the states in the Austro-Germanic Union, who on the occasion we have quoted, officially expressed their desire to line up with these principles? Finally, if we want to extend the principles which will be set down in the general convention for all the European networks and their borders, would it not be fitting to convene the states of Europe's Far East to the conference?[17]

17 «Ne conviendrait-il pas au contraire que tous les Etats de l'Union fussent représentes à Paris […]? Ne conviendrait-il pas, surtout dans l'intérêt des principes qui prévalent maintenant dans l'Europe occidentale, de ne pas négliger l'appui

The Head of French Telegraphs' answer arrived a few days later, but as the Swiss Post Department put it in the report to the Federal Council, it was just "temporary"[18]. Though appreciating the fact that Curchod proposed extending the borders of the union, de Vougny held that he had to discuss the matter with his government, given that the request of his Swiss colleague touched "on very important interests"[19].

Meanwhile the dialogue between Curchod and Brunner went on, with the latter telling the former that a TU without Austria would be unthinkable for at least three reasons. Firstly because the Empire possessed "one of the most widespread telegraph networks"[20] in the whole of Europe and therefore a large portion of telegraph communications risked being excluded. Secondly, following up one of the arguments that Curchod used with De Vougny, Brunner wrote that the Austrian telegraph system was indispensable for the functioning of the European one because Austria was a gateway not only towards Eastern Europe but also towards the Far East and countries like India and China. Finally, Austria would never accept any decisions (on tariffs, telegraph lines, etc.) taken in its absence and without its consent, and that could have serious consequences on the outcome of the conference. Brunner felt that the various states might well have to make "concessions on both sides"[21] (a concept close to his heart since his days as Swiss Telegraphs Head) and without Austria these diplomatic procedures would end in a checkmate.

des Etats de l'Union austro-allemande qui, dans l'occasion que nous avons citée, ont exprimé officiellement leur désir de se ranger à ces principes? Enfin, si l'on veut arriver le plus tôt possible à étendre les principes qui seront poses dans la convention générale à tout le réseau télégraphique européen et à ses aboutissants, ne conviendrait-il pas aussi de convoquer aux conférences les Etats de l'Extrême Orient de l'Europe?». *Ibidem.*

18 «reponse provisoire». *Ibidem.*

19 «touche à des intérêts si importants». *Lettre écrite par de Vougny adressée à Curchod*, 18 Octobre 1864, in SFA, E52, 502.

20 «une des réseaux télégraphiques les plus étendus». *Lettre écrite par Brunner adressée à Curchod*, 10 Octobre 1864, in SFA, E52, 502.

21 «concessions des deux cotés». *Ibidem.*

In November, realizing that the situation was unlikely to be resolved without political intervention, Naeff took the issue to the Federal Council. He tried to get it moving and made an observation that reveals in depth the true reasons behind Switzerland's battle to include Austria in the upcoming European Union. One was that from a geopolitical point of view the TU would be weighted too much towards France. With Austria out of the game, France would be geographically at the centre of the European network and Switzerland would be reduced to a peripheral location in the network and so potentially in political isolation[22]. On the contrary, if the new telegraphic network included Austria, Switzerland were at the centre and income from transit telegraphic traffic increased (of great interest to Switzerland) as would its political weight[23]. It was a precise Swiss doctrine in international diplomacy applied to the field of telecommunications. From the 1850s onwards one of the main aims of Swiss diplomacy was indeed to balance the power between the surrounding countries, and a telegraphic union without Austria would throw these relations out of kilter, and come down in favour of France[24].

A few days later, on 18 November, once again Switzerland made similar moves about relations between technicians and politicians. Curchod wrote to Brunner and informed him of France's interlocutory answer but reassuring him at the same time that Switzerland was ready to put pressure on France "via diplomatic channels"[25]. In fact, on the same day, the Federal Council instructed its ambassador in Paris, the diplomat Johann Konrad Kern, to put before the French government "the proposal to invite the States contracting the Friedrichshafen Agreement, offering, if necessary, its diplomatic support"[26].

22 *Le Département des Postes au Conseil Fédérale*, 10 Novembre 1864, in SFA, E52, 502 and Clark, *International Communications*, 93.

23 *Ibidem.*

24 On the history of Swiss diplomatic strategies in the nineteenth and twentieth centuries see Altermatt, "On Special Mission," 331.

25 «par la vie diplomatique». *Lettre écrite par Curchod adressée à Brunner*, 18 Novembre 1864, in SFA, E52, 502.

26 «la proposition d'inviter les Etats contractants du traité de Friedrichshafen, en offrant, si nécessaire, son appui diplomatique». *Mémorandum par le Conseil Fédérale adressée à Kern*, 18 Novembre 1864, in SFA, E52, 502.

Only ten days later Kern wrote back that without him having to intervene, the French Postal Department had accepted the Swiss proposal. At the Paris Conference the following year all the states that had signed the Friedrichshafen Convention would be invited. Besides approving "thoroughly of the Federal Post Department's point of view", the French government invested the Federal Council with the role of intermediary with Austria[27], given the still uneasy relations between them.

All things considered, Switzerland emerged from this delicate issue, which had so endangered the conference, with a victory right across the board. Firstly, it had obtained the invitation for Austria (with all the geo-political significance it bore for Switzerland). Secondly, it found itself being invested by France with a central diplomatic role because its task would be to smooth things over with Austria, were there to be problems. Finally, though Switzerland had shown itself willing, it had not even had to intervene diplomatically and therefore not expose itself politically[28].

In the early days of December 1864 Curchod wrote again to Brunner and informed him of the positive outcome of the negotiations, hoping that "this preliminary question which I deem so important will thus be happily resolved"[29]. A few days later, Brunner thanked Curchod for his help and emphasized how the successful outcome of the event was basically due to Switzerland's efforts:

> The communication you sent me in your last letter signals a major step towards the completion of a European need and the *honour of the initiative lies with your administration.* I have no doubts about my Government accepting the invitation [...] I thank you for your interest [...] in a question which touches my administration so closely and the interesting communications you have kindly sent me about the subject.[30]

---

27 «entièrement la manière de voir du Département fédéral des postes». *Lettre écrite par Kern adressée au Conseil Fédérale*, 28 Novembre 1864, in SFA, E52, 502.

28 Grossi, "Technologie et diplomatie," 299.

29 «cette question préliminaire à laquelle j'attachais une grande importance se trouvera ainsi heureusement résolue». *Lettre écrite par Curchod adressée à Brunner*, 3 Décembre 1864, in SFA, E52, 502.

30 «La communication que vous me faites dans votre dernière lettre, signale un grand pas vers le complément d'un besoin européen, et *l'honneur de l'initiative en revient à votre administration.* Je ne doute pas que mon Gouvernement accepte

Without such an intervention, both discreet but decisive, by Switzerland, France would have been most unlikely to have invited Austria to the first TU conference. Yet, as illustrated above, without Austria's presence, the new union would have been unlikely to have had the efficacy, capacity to standardize and extend to European telegraph communications that it actually had. The result was that on this occasion as on many others during the 1850s the utopian and avant-garde project to unite European communications would not have been carried out without the intervention of Switzerland.

3.2 The Federal Council calls the tune

In February 1865 the Federal Council gave Kern and Curchod, their two official representatives at the Conference, precise instructions on the aims that Switzerland was to pursue.

Firstly, the delegates were given the mandate to promote the most general and widespread final treaty possible including "as many as possible of the European states"[31].

Secondly, the Federal Council called for all efforts to be made to replace tariff zones with a uniform charge. The fundamental difference was of course that tariffs are uniform when any correspondence between two states is charged at the same rate, while flat rates are applied when the same tariff is paid for any destination. The latter formula was introduced for telegraphs much later and the Federal Council's aim was to replace the zone charge with a flat rate. With the zone system

l'invitation […]. Je vous remercie de l'intérêt […] à une question que touche de si près mon administration, et des communications intéressantes que vous avez bien voulu me faire à ce sujet». *Lettre écrite par Brunner adressée à Curchod*, 9 Décembre 1864, in AF, Fond E52, Archiv nr. 502, [*our italics*].

31 « le plus grand nombre possible de Etats de l'Europe». *Instructions donné par le Conseil Fédéral Suisse à ses délégués Kern et Curchod*, 22 Février 1865, in SFA, E52, 503.

the charge was determined by the distance separating the post office of the sender and that of the receiver, so that it varied according to the town of departure and arrival, and not according to the nation the cities were in[32].

A third and crucial aspect that Kern and Curchod would have to concentrate on was "all combinations to favour telegraphic traffic, especially between neighbouring states, with an application of moderate tariffs"[33]. Stimulus measures would have to be associated with a more underground strategy and explained in detail to the delegates: they would have to be "particularly careful that the dispositions adopted favour as far as possible transit through Switzerland"[34]. This is perhaps the document which states most explicitly that the Swiss interest in a generalized tariff reduction was not only ideological-liberal (i.e. spreading the use of the new means as far as possible), but also and above all for economic reasons. The two delegates would have to move so that the highest possible number of international telegraphs crossed Swiss territory and drew in a hefty income from transit tariffs.

Lastly, the Federal Council invited its delegates to adopt all the necessary provisions "to bring all European telegraph administrations under the one and same regulation", to promote "the most liberal and favourable dispositions for traffic in general", without "losing sight however of the interests of revenue"[35].

---

32 Fari, *Una penisola in comunicazione*, 441.

33 «d'appliquer toutes les combinaisons de nature à favoriser le trafic télégraphique, surtout entre Etats voisins, telles que l'application de taxes modérés». *Instructions donné par le Conseil Fédéral Suisse à ses délégués Kern et Curchod.*

34 «Ils veillèrent en particulier à ce que les dispositions adoptées favorisent autant que possible le transit par la Suisse». *Ibidem.*

35 «rallier dans une seule et même règle toutes les administrations télégraphiques de l'Europe, «les dispositions les plus libérales et les plus favorables au trafic en général, sans perdre de vue cependant les intérêts du fisc». *Instructions donné par le Conseil Fédéral Suisse à ses délégués Kern et Curchod*, 22 Février 1865, in SFA, E52, 503.

3.3 Louis Curchod emerging

The Paris International Telegraph Conference opened on 1 March, in the presence of most of the European countries which at that time possessed a state-run telegraph: Portugal, Spain, France, Belgium, Holland, Sweden/Norway, Switzerland, Italy, Prussia, Hamburg, Bavaria, Austria, Greece and Russia. Each nation was represented by a diplomatic delegate, normally the ambassador stationed in Paris, and by an expert, normally the head of telegraphs or a top manager. The dual delegation was necessary, given that the diplomatic as a representative of his state could sign international treaties while knowing nothing about technical and administrative questions. On the contrary, the technical delegate knew the running details of the service, but did not belong to the diplomatic corps and could not sign conventions which had the value of international treaties. Thus a dual representation guaranteed a high level of technical competence and at the same time the legal value of international treaties.

The French Foreign Secretary Édouard Drouyn de Lhuys was given the task of presiding the conference and in the opening sessions listed the main objectives to be followed during the meeting. A first aim was to improve the conventions stipulated in previous conferences because with the continuous technical progress in telegraphy, some dispositions "were no longer in harmony with the needs and conditions of the present situation"[36]. Another aim was for the various states to commit to standardizing as far as possible European telegraph traffic, having already experimented "the advantages of a uniform telegraph regime for international relations"[37]. A third aim, fundamental for the development of European telegraphy, would be to replace the tariff zones by a uniform rate "in order to simplify as far as possible the use of telegraph

36 «n'étaient plus en harmonie avec les besoins et les conditions de la situation actuelle». *Documents diplomatiques de la conférence télégraphique internationale de Paris*, Imprimerie impériale, Paris 1865, p. 76.

37 «les avantages de l'uniformité du régime télégraphique pour les relations internationales». *Ibidem.*

communications"[38]. In its instructions, the Federal Council had drawn close to sharing a line of thought common to all the other states at least over the three aims to reach during the Paris Conference: technical standardisation, normative uniformity and the abolition of the tariff zones.

Figure 10: The technical delegates at the First Telegraph Union Conference, Paris 1865. *Source: ITU Archives.*

The discussion of the convention articles was left to a commission of telegraph technicians and experts. Some of the most famous scholars and heads of public European telegraph networks took part and among them stood out the figures of the Frenchman Vougy, the Austrian Brunner von Wattenwyl, the Belgians Fassiaux and Vinchent, the Italian Minotti, the Prussian De Chauvin and, of course, Curchod. What

38 «afin de simplifier autant que possible l'emploi des communications télégraphiques». *Ibidem*, p. 77.

happened was that for the first time a large group of technicians with a long experience in telegraphs all met. It was the coming together of the European "technical elite" which from the 1850s onwards had interwoven bilateral and multilateral relations and was now deciding on the destiny and evolution of international telegraphy[39]. Furthermore, following the Swiss proposal, each state had the right to one sole vote, irrespective of the number of delegates in its commission[40]. It was a profoundly democratic mechanism allowing each country the same weight and possibility of influencing others, whatever the size of its population.

Although the pre-eminence which was to characterize the Swiss participation in the 1868 Vienna Conference was not yet evident, Curchod knew how to move with intelligence and equilibrium during the long debates. Thanks to his long experience in running of a telegraph service, more or less since about the introduction of telegraphy in Europe, he enjoyed great respect among the other technical delegates. It was probably this respect, at times almost reverence, which allowed Curchod to intervene in almost all the important discussions, reserving for himself the last word before the approbation or rebuttal of an amendment. In this subtle game of balancing the equilibriums within the conference, he had to contend with the equally competent Vinchent and with Jagerschmidt, the French host and as such president of the technical sessions. Differently, however, from Vinchent and Jagerschmidt, Curchod was able to impose his line of thought in most circumstances. For example, he convinced the delegates to converge on his position about suppressing urgent telegrams, recorded deliveries, telegram lengths and charges, and finally he drew up with Brunner a detailed table indicating tariffs for correspondence between the different countries. Curchod managed to obtain these results with a dual strategy. While imposing his line of thought by appealing to his famed

39 On the technical and political ability of this elite see Laborie, *L'Europe mise en réseaux*. For a different approach see S. Fari, G. Balbi and G. Richeri, "Last in First Out. How the Telegraph Brought to International Organization before Post and Trains" (paper presented at the International Committee for the History of Technology Conference, Barcelona, July 2012).

40 *Documents diplomatiques de la conférence télégraphique internationale de Paris*, p. 78.

experience, he often mediated between the positions of countries with diametrically different ideas[41].

His ability in manoeuvring was also shown in the way he worked towards the main objectives of his country. A first result had been reached before the opening of the conference. Thanks to the contribution of the Swiss Post Department and Curchod himself, Austria also took part in the first TU Conference, thus preparing ideal grounds for arriving at a European treaty which was as general as possible.

In Paris Curchod's role was determining especially about tariffs, the most sensitive subject treated during the conference. On the shift from zone-based to uniform tariffs, Switzerland had to mediate between a Spanish proposal, surprisingly in favour of small states like Switzerland, and the legitimate needs of geographically far larger countries. The Spanish delegate Hakar suggested that "in all countries the same number of words will have the same tariff, so that the cost will multiply as many times as there are countries to cross"[42]. In practice the project proposed the same tariff for a telegram to and from any country, which implied the adoption of a flat rate for the whole of Europe. Most other countries including France, however, suggested introducing a uniform tariff between country and country, substantially each state would establish its own tariffs for communicating with another country. Faced with the Franco-Belgian attack and a Spanish project, Curchod took the word and indicated "how he would understand the proposition of the Spanish delegates"[43]. Curchod did not intervene in favour of the Spanish amendment but:

> proposed an amendment with the following bases: 1) the number of territorial divisions where a separate charge will be made would not be more than two per state; 2) the charge would vary between a minimum of fifty centimes and a fixed

41 This mediatory work was not undertaken by Curchod alone, but also by many other delegates who wanted the success of the Conference, like the Belgian, French and Italian directors.

42 «dans tous les pays, la même nombre de mots aura la même taxe, de sorte que la taxe devra se multiplier autant de fois qu'il y aura de pays à traverser». *Ibidem*, p. 151.

43 «comment il entendrait la proposition de MM. les Délégués de l'Espagne». *Ibidem*, p. 152.

maximum; 3) the monetary base would be the franc, and no lower fractions than fifty centimes would be allowed.[44]

In this way he reached four purposes: allowing the great empires (Russia and Austria) to keep a double tariff, justified by their extensive territories; fixing a uniform tariff count; respecting the liberal principle of leaving every state to determine the exact price, thanks to an oscillating tariff range; pleasing France by putting the French franc as the yard measure. To show the excellence of his proposals Curchod emphasized that this was the *modus operandi* in Switzerland, signalling in this way the high level of his administration and nation, after which "he would indicate the objective to be reached by the Conference. It had to, according to him: 1) decide on the monetary base of a simple telegram; 2) indicate the method to be used for establishing the tariffs between various states; 3) fix the maximum number of territorial divisions corresponding to distinct tariffs"[45]. His indications not surprisingly were followed almost to the letter during the whole discussion over the question of tariffs.

The prolonged debate over tariffs induced Curchod to formalise his suggestions in new regulations for articles 30 and 31, which were unanimously approved and inserted in the text of the convention. Apparently it did not appear substantially very different from the French proposal, though actually differing in some fixed points: the introduction of a uniform tariff "applicable to all correspondence exchanged along the same route between the offices of any two contracting states"[46]; senders would be free to select the route for their

44 «Il propose un amendement dont les bases sont les suivantes: 1) le nombre des divisions territoriales auxquelles s'appliquerait une taxe distincte ne dépasserait pas deux par État; 2) la taxe varierait entre un minimum de cinquante centimes et un maximum déterminé; 3) la base monétaire serait le franc, et l'on n'admettrait point de fraction inférieur á cinquante centimes». *Ibidem.*

45 «il indique le but á atteindre par la Conférence. Elle doit, selon lui: 1° déterminer la base monétaire de la dépêche simple; 2° indiquer la manière dont se fera l'établissement des tarifs entre les divers États; 3° Fixer le nombre maximum des dimensions territoriales auxquelles correspondront des taxes distinctes». *Ibidem.*

46 «la taxe applicable à toutes les correspondances échangées par la même voie entre les bureaux de deux quelconques des États contractants sera uniforme». *Ibidem.*

telegrams and therefore choose between various price options; the twenty word limit for simple telegrams; the possibility to keep two different tariffs within very big countries like Russia and Austria; finally the French gold franc as currency.

Probably in order not to be accused of being illiberal and at the same time ward off France, Curchod ended by establishing that "the tariff rates will be established from state to state, in accord with the governments both at intermediate points and destiny"[47]. To round off he explained very clearly the reasons for the small adjustments that his final amendment showed in comparison to his initial proposal:

> He put on the front line as the fundamental basis of the treaty the principle of territorial unity, and gave only second place to the exception, which is only a sacrifice of the freedom of action of the big states. He had to introduce in article 30 the definition of a simple telegram, to arrive at formulating in a precise way the dispositions for monetary unity, which he highly recommends adopting.[48]

What mattered to Curchod was to obtain uniformity not so much in the tariffs as in the overall running of the telegraph network. This meant establishing a minimum cost for single telegrams, decide on the currency for the calculations and above all indicate tariffs between states in a single comprehensive table.

If the Spanish project for uniform tariffs was rejected, it was also because it may have seemed in favour of small states, but in reality it contrasted with another basic Swiss interest – transit traffic. Curchod insisted on this, declaring:

47 «le taux de la taxe sera établi d'État á État, de concert entre les Gouvernements extrêmes et les Gouvernements intermédiaires». *Ibidem.*

48 «Il a placé en première ligne, comme étant la base fondamentale du traité, le principe de l'unité territoriale, et n'a donné que la seconde place à l'exception, qui n'est qu'un sacrifice à la liberté d'action des grands États. Il a dû introduire dans l'article 30 la définition de la dépêche simple, pour arriver à formuler d'une manière précise les dispositions relatives à l'unité de monnaie, dont il recommande tout particulièrement l'adoption». *Ibidem*, p. 159.

> there's no problem in the charges varying from one country to another according to the route followed, and that a telegram from Austria to Spain, for example, can be charged differently, depending on it passing through Switzerland or Italy.[49]

Switzerland's aim, which seemed, however, to be shared with most of the other countries, was that transit tariffs needed to be kept low to encourage international traffic. But differently from Belgium, Austria and Sweden, who suggested regulating at least some parameters globally, Switzerland was interested in leaving complete freedom of decision to each country, in other words leaving the tariffs open to free competition.

Above all, since the international telegraph network was extremely ramified and a telegram going between two non neighbouring countries could use different routes, a precise criterion was established which telegraph operators had to observe. When sending the telegram they were to follow the cheapest route, and not as had first been proposed, the most direct[50]. This change fitted in with the liberal attitude which had been followed during the whole conference. The choice of the cheapest route guaranteed free competition among the various European administrations running the international service. Except for the case of senders indicating otherwise, telegraph operators were to direct the telegram where the tariff was lowest, the aim being to stimulate competition in constructing cheaper routes.

Once the general principles for international tariffs had been established, based mostly on his "suggestions" and "adjustments", Curchod went on to a second aspect of tariffs, with the drawing up of a general table of transit and destiny tariffs valid for all member states. Initially the debate was taken up by a discussion over the method to adopt for drawing up the table. Some, like the French delegate and Curchod himself, thought it appropriate to keep to the liberal principle and leave each state to decide bilaterally with its neighbours transit and destination tariffs,

49 «il n'y a aucun inconvenant à ce que les taxes d'un pays à un autre varient selon la voie sui vie, et qu'une dépêche de l'Autriche pour l'Espagne, par exemple, puisse être taxée différemment, selon qu'elle devra passer par la Suisse ou par l'Italie». *Ibidem*.

50 «voie moins coûteuse», «plus directe». *Ibidem*, p. 189.

also by means of post-Conference conventions. The Belgian Vinchent, instead, insisted on a tariff table being appended to the Convention. To settle the diatribe, the chairman of the session Jagerschmidt put to the vote a question worthy to be put before Solomon: "Will there be tariffs in the International Convention project?"[51].

The delegates voted yes, so that the dilemma opened on how to elaborate such a tariff table and it was agreed for it to be drawn up at a later stage. Only after a lengthy discussion involving some of the leading figures (Curchod, Faisseux the second Belgian delegate and Jagerschmidt), did the tariff issue come to its conclusion:

> After a short exchange of comments between the President, Messrs Faisseux and Curchod, it was agreed that without interrupting Commission's work in progress, the delegates would deal, first of all individually and then together, with preparing the table of charges which in the terms of Colonel de Chauvin's proposal must appear as an attachment to the treaty.[52]

On the same occasion it was emphasized that in line with liberal principles the tariffs inserted in the table would only be general indications, under which the various states could drop as much as and when they so desired [53]. As mentioned above, the debate finished with the presentation of two tables drawn up by Brunner and Curchod together.

In the definitive version, Switzerland adopted the lowest transit tariff (one French franc) valid also for all correspondence transiting through the country. It meant that after the Paris Conference sending transit telegrams over Swiss territory would cost appreciably less than elsewhere and therefore telegraph operators would be obliged to use that route. One of the main aims of the Federal Council (guaranteeing and increasing transit telegraph traffic, so important for the entire running of the Swiss telegraphs) had been reached and the country would find itself in

51 «Y aura-t-il un tarif dans le Projet de Convention internationale?». *Ibidem*, p. 175.

52 «Après un court échange d'observations entre M. Le Président, MM. Faissaux et Curchod, il est convenu que sans interrompre le cours des travaux de la Commission, MM. les délégués s'occuperont, d'abord individuellement, puis de concert, de la préparation du tableau des taxes qui doit, aux termes de la proposition de M. le colonel de Chauvin, figurer comme annexe au traité». *Ibidem*, p. 187.

53 Art. 31, third section of the Paris Telegraph Convention 1865.

a privileged position in the European context. The great importance of the tariff reduction for Switzerland in the Paris Conference was recalled by Curchod in an interview during the 1868 Vienna Conference. In order to defend article 31, which established that the tariffs in the general table could always be reduced, Curchod stated expressly:

> If the amendment system had prevailed in 1865, Switzerland would have had to give up the whole of its part in the transit between western and eastern Europe; for it is only by reductions on Convention charges, operated in agreement with Austria that it has been able to attract to its lines both the correspondence exchanged between France and Austria and that of the same origin for Turkey, Russia etc. The Swiss Office has not profited only from these combinations; they have also gone to the advantage of the public in opening up multiple routes. M. Curchod is however of the general opinion that it is only by the nature of the service and not by exaggerated reductions in charges that the states must fight among themselves and while maintaining the principle of liberty, he would refuse to introduce a restriction in this sense.[54]

Three other important matters were discussed at the Conference without any active participation from Curchod, perhaps because the Federal Council had expressed no opinion about them: a method for running the accounting of telegrams between bordering states; the frequency of general conferences; finally the creation of a permanent body to run international traffic.

The issue of settling credits and debts between neighbouring states was very important. In fact, the country from which the telegrams left

54 «Si le système de l'amendement eut prévalu en 1865, la Suisse aurait dû renoncer à toute part dans le transit entre l'Europe occidentale et l'Europe Orientale; car ce n'est que par des réductions sur les taxes de la Convention, opérées de concert avec l'Autriche, qu'elle a pu attirer sur ses lignes, soit les correspondances échangées entre la France et l'Autriche, soit celles de même origine pour la Turquie, la Russie, etc. L'office Suisse n'a pas seul profité de ces combinaisons ; elles ont encore tourné à l'avantage du public, en ouvrant à ses correspondances des voies multiples. M. Curchod partage toutefois l'avis général que c'est par la nature du service et non par des réductions de taxes exagérées que les Etats doivent lutter entre eux, et, tout en maintenant le principe de liberté, il ne refuserait pas d'y introduire une restriction dans ce sens». *Documents de la conférence télégraphique internationale de Vienne*, Imprimerie impériale et royale de la Cour et de l'Etat, Vienne 1868, p. 308.

took in payment for whole transmission, irrespective of the number of countries the telegram had to cross. Obviously with the thousands of telegrams crossing borders every month, the choice of a good accounting system was essential, in part to avoid any disparity or ill-feeling. At the same time the accounting system needed to be able to calculate swiftly and as accurately as possible the sum of reciprocal credits and debts without having to count payments one by one. Vinchent proposed adopting a method already operating in the more advanced administrations, including the Belgian one, and called the "averages system":

> Each state credits the neighbouring state with total amount of the charges for all the telegrams it sends, calculated from the border between the two states up to destination. These charges can be settled according to agreement after the number of telegrams franked at the border, leaving out the number of words and extra expenses. In this case the parts of the neighbouring state and each of the following states, if involved, are determined by the averages established in a contradictory mode.[55]

The conference did not accept the criterion as compulsory, but at the end of the debate decided to insert it as the method advised by the Convention which the states could implement at will. For the second question a Belgian suggestion was again accepted, and it was established that at the close of a conference the place and date of the next would be established (art. 55). Finally, Sweden and Austria spoke against a permanent body to run the international telegraph service, holding it to be useless as well as difficult to convene all the participants every time. France, which had put forward the suggestion, replied as follows:

> that is not the sense of the article: the Commission proposed by the Project *would be in effect permanent*, in the sense that it will have to accomplish a continuous mission; but this mission could be carried out *without the presence of delegates from each state having to be present*. Periodic meetings would be sufficient, and

55 «chaque État crédite l'État limitrophe du montant des taxes de toutes les dépêches qu'il lui a transmises, calculées depuis la frontière de ces deux États jusqu'à destination. Ces taxes peuvent être réglées de commun accord d'après le nombre de dépêches qui ont franchi cette frontière, abstraction faite du nombre de mots et des frais accessoires. Dans ce cas, les parts de l'État limitrophe et de chacun des États suivants, s'il y a lieu, sont déterminées par des moyennes établies contradictoirement». *Ibidem*, p. 144.

even for these meetings, the different groups of states could come to agreements and entrust a common mandate to one delegate.[56]

As said above, Switzerland expressed no opinion and the suggestion was rejected without further deliberation. A permanent office for the day-by-day management of international traffic (what would be called the Bureau) was established in the 1868 Vienna Conference, this time following a suggestion from Switzerland and realizing a project of Curchod's.

3.4 Switzerland scores

The Federal Council was fully satisfied with the outcome of the Conference and the work of Kern, their Parisian ambassador, and Curchod.

In the first place the newly signed convention was presented to Parliament as the full realisation of a closely followed national project. The Paris treaty functioned as a way out of a "moment of transition"[57] made of bilateral treaties and various conventions signed by the two preceding unions. It was "a general international treaty extending advantages obtained to all European correspondence", but felt to be of prime importance to Switzerland ("we did everything in our power to reach this objective"[58]).

Besides bringing into being a telegraph convention common to all Europe, the Federal Council cited three other relevant cases of progress that standardization and uniformity had added to international

56 «tel n'est pas le sens de l'article: la Commission proposée par le Project *serait en effet permanente*, en ce sens qu'elle aurait à accomplir une *mission continue*; mais cette mission pourrait être remplie *sans que la présence de délègues de chaque Etats fut toujours nécessaire*. De réunions périodiques y suffiraient, et, pour ces réunions même, les divers groupes d'Etats pourraient s'entendre et confier un mandat commun à un seul délègue». *Ibidem*, p. 216.

57 «un état transitorie». FF, 3/33, 22 Juillet 1865, p. 131.

58 «un traité international général étendant à toutes les correspondances européennes les avantages obtenus», «nous pousserions de tout notre pouvoir à la réalisation de ce but». *Ibidem*.

telegraphy. Firstly, it had been established that "all the languages spoken on the territory of the contracting states" could be used for telegrams. Secondly, the French franc was chosen as the only currency for calculating the tariffs. Finally, it was emphasized that "a uniform tariff between the offices of any two of the contracting States (art. 30), and a considerable reduction of the present tariffs, as the consequence of applying the said principle"[59].

Among the positive outcomes of the Paris Convention for Switzerland, the Federal Council concentrated principally on the new tariffs. According to article 31, a tariff could be "always and in any moment reduced by a common agreement between such and such interested governments"[60] was for Switzerland a guarantee that it could continue to follow its tariff containment policy. The TU constitution allowed single states the power to establish reciprocal conventions and among other things to modify tariffs. Switzerland applied this article to go on reducing tariffs between country and country and for transit, so encouraging telegraphic communication between itself and Europe as well as favouring international telegrams over its territory.

The generalized lowering of tariffs carried out in Paris would give Switzerland a second advantage, substantially increasing telegraph traffic especially towards countries with which there was intense commercial exchange. For example, thanks to a convention made with Italy, Switzerland would be able to exchange telegrams with many of the northern cities at the cost of 2 francs instead of 3, though to obtain this privilege it would have to give up the 1.50 franc tariff which up to then was charged at the border offices. Substantially

> To give only one example, Bellinzona corresponds now for 1.50 francs with Como but it pays 3 francs for Milan or Turin; henceforth it will pay 2 francs for each of the three cities; now its traffic with Como will always be of little importance in

59 «toutes les langues usitées sur le territoire des Etats contractants», «une taxe uniforme entre les bureaux de deux quelconques des Etats contractants (art. 30), et la réduction considérable des taxes actuelles qui a été la conséquence de l'application de ce principe». *Ibidem,* p. 132.

60 «toujours et à toute époque être réduite d'un commun accord entre tel ou tel des Gouvernements intéressés». Ibidem.

comparison to that with Milan and Turin; it will earn in proportion appreciably and also extra.[61]

A page 139.

Tableau comparatif

des taxes télégraphiques actuelles et futures.

Taxes des dépêches de vingt mots échangées entre les bureaux suisses de :

	1. Beaurivage, Genève, Lausanne, Montreux, Sion. situés dans la 1re zône de la frontière franco-suisse. 2e zône de la frontière suisse austro-allemande.				2. Aarau, Bâle, Berne, Berthoud, Chaux-de-Fonds, Frauenfeld, Fribourg, Lucerne, Neuchâtel, Schaffhouse, Soleure, Winterthour, Zofingen, Zurich. situés dans la 1re zône de la frontière franco-suisse. 1re zône de la frontière suisse austro-allemande.				3. Appenzell, Coire, Flawyl, Glaris, St. Gall, Schwyz, Trogen. situés dans la 2e zône de la frontière franco-suisse. 1re zône de la frontière suisse austro-allemande.				4. Andermatt, Bellinzone, Chiasso, Dissentis, Interlaken, Locarno, Louëche-les-Bains. situés dans la 2e zône de la frontière franco-suisse. 2e zône de la frontière suisse austro-allemande.			
Bureaux étrangers.	Taxe.				Taxe.				Taxe.				Taxe.			
	Actuelle.		Future.		Actuelle.		Future.		Actuelle.		Future.		Actuelle.		Future.	
	Fr.	Ct.	Fr.	Ct.	Fr.	Ct.	Fr.	Ct.	Fr.	Ct.	Fr.	Ct.	Fr.	Ct.	Fr.	Ct.
Autriche. Trieste, Vienne	5	—	4	—	4	—	4	—	4	—	4	—	5	—	4	—
» Insbruck	4	—	2	—	3	—	2	—	3	—	3	—	4	—	2	—
» Venise	4	—	3	—	3	—	3	—	3	—	3	—	4	—	3	—
Bade : Carlsruhe	2	—	2	—	2	—	2	—	2	—	2	—	2	—	2	—
Bavière : Munich	2	—	2	—	2	—	2	—	2	—	2	—	2	—	2	—
Belgique : Bruxelles	7	50	4	—	7	50	4	—	9	—	4	—	9	—	4	—
» Anvers	9	—	4	—	9	—	4	—	10	50	4	—	10	50	4	—
Danemark : Copenhagen	15	—	5	50	13	50	5	50	13	50	5	50	15	—	5	50
Espagne : Cadix	15	—	6	50	15	—	6	50	16	50	6	50	16	50	6	50
» Madrid	12	—	6	50	12	—	6	50	13	50	6	50	13	50	6	50
» Barcelone	10	50	6	50	10	50	6	50	12	—	6	50	12	—	6	50
France : Hâvre, Marseille, Paris, etc., (uniforme)	3	—	3	—	3	—	3	—	3	—	3	—	3	—	3	—
Grèce : Athènes	19	50	8	—	18	—	8	—	18	—	8	—	19	50	8	—
Italie : Milan, Turin, etc.	3	—	2	—	3	—	2	—	3	—	2	—	3	—	2	—
Florence, Naples, Palerme, etc.	3	—	3	—	3	—	3	—	3	—	3	—	3	—	3	—
Pays-Bas : Amsterdam	5	—	4	—	4	—	4	—	4	—	4	—	5	—	4	—
Portugal : Lisbonne	16	50	7	50	16	50	7	50	18	—	7	50	18	—	7	50
Prusse : Berlin	5	—	4	—	4	—	4	—	4	—	4	—	5	—	4	—
Russie : St. Pétersbourg	21	—	9	—	19	50	9	—	19	50	9	—	21	—	9	—
Suède : Stockholm	20	—	7	—	21	50	7	—	21	50	7	—	23	—	7	—
Norvège : Christiania	24	50	9	—	23	—	9	—	23	—	9	—	24	50	9	—
Turquie : Constantinople	19	50	8	—	18	—	8	—	18	—	8	—	19	50	8	—
Wurtemberg : Stuttgart	2	—	2	—	2	—	2	—	2	—	2	—	2	—	2	—

Figure 11: Table comparing telegraph tariffs before and after the 1865 Paris Conference. Tariffs for 20 word maximum telegrams exchanged between Swiss telegraph offices and various countries.

Source: FF, 3/33, 22 Juillet 1865, Message du CF à Assemblée fédérale concernant le traité télégraphique international de Paris, *p. 139.*

61 « Pour ne citer qu›un exemple, Bellinzona correspond maintenant pour fr. 1.50 avec Como, mais il paie fr. 3 pour Milan ou pour Turin; dorénavant il payera fr. 2 pour chacune de ses trois villes; or son trafic avec Como sera toujours peu important vis-à-vis de celui avec Milan et Turin; il gagnera donc sensiblement à la mesure, et ainsi des autres ». *Ibidem,* pp. 136–137.

The Federal Council attached to this memorandum a table which showed clearly how there were now notable reductions in charges in telegrams sent between the main Swiss cities and other European ones. From a political and economic point of view Switzerland came away from the Paris Conference with closer ties to other European nations.

To achieve these results, however, the Swiss delegates had had to make "reciprocal concessions [...and] keep in mind the interests and very different conditions of the administrations represented"[62]. Perhaps because it was infused by a liberal spirit more than most European nations, Switzerland was aware of the need to come to terms with other countries which perhaps were less developed telegraphically or had different needs or visions.

On the other hand, there were at least two negative consequences Switzerland would have to bear with, due to the introduction of the new tariffs. The Federal Council described them as a "natural consequence of the system of uniform tariffs"[63]. First of all, as the Federal Council informed Parliament very clearly, there would be a drop in income because of telegrams to/from very distant countries and those with which there were no particular commercial relations. In the latter case the considerable reduction in the tariffs established in Paris would not result in a sufficient rise in traffic, given that there was little[64]. However, this drop in income would be more than made up for by the increase in traffic encouraged by the lower tariffs with nearby countries and those with which there were intense trade interests.

Secondly, and there was no compensation for this, the Federal Council foresaw a drastic reduction in income coming from transit telegrams between neighbouring countries. For telegrams Paris/Milan and Berlin/Turin, for example, with the zone tariffs telegrams would cost the same if they transited across Switzerland or went directly via the two

62 «*concessions réciproques* [...], tenir compte des intérêts et des conditions très diverses dans lesquelles se trouvaient les administrations représentées». *Ibidem*, p. 132, [*our italics*].

63 «conséquence naturelle du système des taxes uniformes». *Ibidem*, p. 134.

64 *Ibidem*, pp. 133–134.

countries involved[65]. With the introduction of uniform tariffs, however, telegrams Italy/France and Italy/Germany would cost one more franc than direct dispatch between the two countries. An estimate was given of how much the state coffers would lose. In 1862 Switzerland took in 80,000 francs a year for transit telegrams, which went down to 60,000 in 1864 and was estimated to drop even further to 30,000 in 1866. It is not clear if this calculation includes income from transit telegrams from non-bordering countries.

3.5 Switzerland beefs up

On the whole, despite the concessions made Switzerland came away from the first TU conference strengthened. It was politically stronger because of its central role in inviting Austria, for without the latter the conference could not have had a successful outcome. Switzerland's international prestige was on the up and up, firstly because of the role played by Curchod, who was evidently enjoyed great credibility and was skilled in mediating in the most complicated situations. Its technology and telegraphy were also stronger. With the introduction of the cheapest route and the move to the cheapest transit tariffs Switzerland literally found itself at the centre of the period's international telegraph traffic. To sum up, the Paris Conference laid the basis for Switzerland to be acknowledged as having an important role in the rising sector of European telecommunications, a role which would lead to the country being assigned the running of the Bureau in 1868.

65 From Berlin to Turin the transit was classified as direct because it could use the AGTU networks. FF, 3/33, 22 Juillet 1865, p. 134.

Chapter 4
The 1868 Vienna Conference

Introduction

The International Telegraph Conference was held in Vienna three years after the ratification of the TU in Paris. A brief period, but fraught with social and political turmoil in Europe, mostly due to the slow but ongoing formation of national states that undermined the stability of the multi-ethnic Austrian Empire. The 1866 Austrian-Prussian War was fought between the Austrian Empire, Saxony, Bavaria, Württemberg, Hanover and other minor German states on one side and Prussia, Italy and various lesser German states on the other. There were two major consequences, one being the gradual affirmation of a Prussian-led German Empire destined to become one of the greatest European powers. The second was that thanks to Prussian backing and Napoleon III's mediation, Italy strengthened its national identity by winning the Veneto area[1].

Defeated in war and perceived by the nascent European nationalist movements as an outdated political order, Austria was forced to bring about a radical institutional reform in order to avoid falling apart. This it did by coming to a compromise (*Ausgleich*) with its Hungarian subjects, reflected in renaming itself the Austro-Hungarian Empire[2].

Another big European state, which took part in the first TU Conference in Paris, underwent a bourgeois uprising in the same period. The

1 G. Wawro, *The Austro-Prussian War: Austria's war with Prussia and Italy in 1866* (Cambridge and New York: Cambridge University Press, 1996).

2 M. Cornwall (ed.), *The last years of Austria-Hungary: a multi-national experiment in early twentieth-century Europe* (Exeter: University of Exeter Press, 1990).

Glorious Spanish Revolution, planned in 1866/7 by exiled liberals and republicans, took place between September and December 1868, when Queen Isabella II was forced to abandon the country for France. The result was another autocratic state opening a democratic and liberal parenthesis, known as the Democratic *sexenio* (1868–78), which was then brought to an end by the restoration of the Bourbons[3].

These political facts provided the backdrop for the Vienna Conference, where a relatively stable Switzerland expected to make further progress with the liberalisation of international telecommunications. Curchod, still head of Swiss telegraphs and again technical delegate to the conference, had drawn up a completely different kind of project which would have a permanent influence on the structures of the world telegraph organisation and also on the Swiss position on the international chessboard.

4.1 A special agent for international telegraphy

With a note dated 8 February 1868, the Austrian Embassy in Bern informed the Federal Council that the second International Telegraph Conference would be taking place the following June and that it would have a different format from Paris. The Austrian government wanted to organize a "technicians' reunion"[4] in order to resolve issues that had emerged in the running of the international system and limit as far as possible the presence and importance of diplomats. The Swiss Post Department therefore put forward the name of the Head of Telegraphs, who was immediately appointed by the Federal Council "quite naturally"[5] as

3 G. Thomson, *The birth of modern politics in Spain: democracy, association and revolution, 1854–75* (Basingstoke and New York: Palgrave Macmillan, 2010).

4 «reunion de techniciens». *Le Département des Postes au Conseil Fédérale*, 9 Mars 1868, in SFA, E52, 508.

5 «tout naturellement». *Ibidem*.

the sole Swiss delegate to Vienna. His great technical experience was proven, as was his diplomatic ability.

The day after being officially appointed delegate to the new conference, Curchod sent the Head of the Swiss Post Department Jean-Jacques Challet-Venel a long memorandum detailing objectives he wanted to achieve at Vienna. In particular, he focused on improvements to the everyday running of the international service because there had been some "inconveniences that practice alone could reveal and that consequently were difficult to prevent"[6].

Curchod believed that, for this purpose, articles 31 and 60 of the Paris Convention needed to be both defended and improved. Article 31, which regulated how to establish international tariffs and possible modifications, needed to be kept for purely administrative reasons. Even if new norms risked being delayed, given the need for a consensus over amendments, article 31 as it stood avoided the "misunderstandings, complications and irregularities"[7] which could cause such harm in telegraph management. Article 60 instead regulated adhesion to the convention, which was limited to states applying for it. Curchod insisted on including private companies too, again to fill a legislative gap and avoid complications in the day-to-day management of the system.

These articles are only two examples of what Curchod felt to be the main problem that had emerged in the early years of international telegraphy. What the service needed was an ordinary and regular management which periodic conferences could not guarantee. To remediate the situation he made a deft proposal to the Head of the Post Department, which was only slightly changed before being approved at Vienna and so giving life to the Bureau.

First, Curchod re-examined the French project, which had been so quickly turned down at the Paris conference and identified some of the

6 «inconvénients que la pratique seule devait mettre en évidence et qu'il était par conséquent difficile de prévenir». *Lettre écrite par Curchod adressée à Challet-Venel*, 12 Mars 1868, in SFA, E52, 508. All the following quotations are from the same source.

7 «malentendus, des complications et des irrégularités».

reasons for its failure. Since the "differences of opinion"[8] among the various states were everyday happenings, the need for a body mandated to settle such problems between one conference and another was becoming a pressing reality. While there had been no "systematic opposition"[9] to the creation of a commission, what had aroused doubts was its structure and organisation "with its permanent character and composed of delegates from all the contracting states leaves too great a margin for the unforeseen and the unknown[10]. Briefly, the French-proposed commission did not seem able to face the everyday problems of international telegraphy because it would take too long to convene all the member states and consequently come to any decisions. The alternative was to appoint a special agent chosen by the conference, placed at the dependence of the telegraph administration of the state hosting the upcoming conference and salaried by all member countries.

> This special agent, to whom we will give for example the title of Conference *Secretary General*, would replace for everyday, low-profile running business the Commission discussing him, the Commission which nothing forbids from also convening, but in order to be summoned only when unlikely cases or serious difficulties arise.[11]

In other terms, Curchod did not reject the idea of a special commission, but held that its functions were limited to rare exceptional events disturbing the smooth running of telegraph communications. Differently, the special agent would be dealing with the current everyday aspects concerning the practical application of the convention, apparently less important but fundamental for the running of the service.

---

8 «divergences d'opinion».

9 «opposition systematique».

10 «avec son caractère permanent et composée des délègues de tous les Etats contractants laissait trop de marge à l'imprévu et a l'inconnu».

11 «Cet agent spécial, auquel nous donnerions par exemple le titre de *Secrétaire général* des Conférences, remplacerait pour les affaires courantes et de peu d'importance la Commission dont il vient d'être question, Commission qui rien n'empêche de constituer aussi, mais pour n'être appelée que dans les cas peu probable où une difficulté sérieuse viendrait se présenter». [*Our italics*]. Still today the title of secretary general is used for the Head of the ITU.

First of all, the agent would act as "referee for all matters concerning common interests"[12] and would centralize problems and proposals. It would be up to him to periodically collect and send out to member states information about future administrative and tariff changes. This procedure would bring three main advantages. It would reduce the number of communications between the various states, because they would no longer have to contact all the others one by one. Secondly, attributing to one figure the task of centralizing communications meant fewer chances of neglect or delay. It had happened in fact that some telegraph administrations had seriously harmed the running of the service by not notifying modifications, either through poor judgement or through fear of creating disturbance over apparent trifles. Finally, a third advantage given by the centralization was greater clarity and streamlining data and communications.

The central agent would not only have bonding/referring functions with the member states, but would also represent the union in its relations with non-member bodies. In other words, the newly appointed secretary general would provide an interface and regulate relations between the TU and those nations which had not signed the convention in addition to private telegraph companies.

Finally, the figure profiled by Curchod would also deal with collecting and forwarding to all the states a series of statistics on the functioning of the telegraph service in various countries, especially tariffs.

This solution offered at least three advantages over the French proposal and traditional conferences. The secretary would have a permanent seat while the French project foresaw periodic meetings taking place in different countries "two or three years this service will change hands; on changing hands it will more or less change organisation, it will no longer have any following or system"[13]. Then all the member states would contribute to financing the agent and his office and the expenses would be divided up in equal parts. In appointing a figure in charge of the service states needing to communicate possible variations

12 «référendaire dans toutes les questions touchant aux intérêts communs».

13 «les deux ou trois ans ce service changera de main; en changeant de main il changera aussi plus ou moins d'organisation, il n'y aura plus ni suite, ni système».

"would not fear troubling"[14] and would feel free to act. Curchod insisted in particular on the practical need to set up a permanent seat for the new international agency. How much the actual organisation of the Federal Council influenced Curchod's reasoning can be seen in the comparison he made with the pre-1848 Swiss system of institutions:

> 25 years ago, the management of federal affairs was in the hands of a canton government and passed every two years from one government to another. *The government changed, together with its place of residence, but what did not change was the Federal Chancellery, armed with its archives, documents and traditions,* which prepared the work for the Vorort[15] and implemented its decisions. Thanks to the Federal Chancellery, these regular changes of government took place with no upsets, no business coming to a halt, no compromising the unity of the system and the canton governments could take charge and run affairs without becoming vulnerable to heavy sacrifices, for Chancellery expenses were covered by all the Cantons together.[16]

According to Curchod, some institutions like the plenary conferences or the French commission needed to have an itinerant seat for essentially political reasons, just as the Swiss government moved its residence every two years from canton to canton. The special agent dealing with international telegraph management needed to have a permanent site like the Federal Chancellery. Space was needed for archiving documentation and a stable bureaucratic apparatus was essential. Furthermore, it was the case of introducing one of the strong points of the Swiss political model, federalism, into the running of the European telegraph

14 «craindraient d'importuner».

15 The term used for Federal Council/Government in the period.

16 «dans la Confédération Suisse, il y a 25 ans, la direction des affaires fédérales était entre mains du Gouvernement d'un Canton et passait tous les deux ans d'un Gouvernement à l'autre. *Le Gouvernement changeait ainsi que le lieu de résidence, mais ce qui ne changeait pas c'était la Chancellerie fédérale, armée de ses archives, de ses documents et de ses traditions,* qui préparait le travail du Gouvernement du Vorort et mettait ses décisions à exécution. Grâce à la Chancellerie fédérale ces changements réguliers de Gouvernement se faisaient sans choc, sans arrêt des affaires sans compromettre l'unité du système et les gouvernements des Cantons directeurs pouvaient d'un autre coté se charger successivement de la direction des affaires fédérales sans s'exposer à des sacrifices sensibles, car les frais de la Chancellerie étaient couverts par l'ensemble des Cantons». [*Our italics*].

network. It was not the first time that the Swiss model of "unity in difference" had been indicated as applicable on a wider European scale. Already in the 1850s Switzerland had been seen as a possible solution for a future European unity[17].

After outlining in detail the figure of the secretary and the utility and structure of the Bureau, Curchod brought his long letter to Challet-Venel to an end by emphasizing two interesting aspects. Firstly, perhaps to reaffirm the freedom and sovereignty of the single countries also in terms of telegraphy, he recalled that the agent would not at all be an "indispensable middleman"[18] in communications between administrations, but on the contrary each country would be free to establish private agreements with others. Secondly, Curchod also recalled that once the TU had "an official and recognized body", it would exercise a "heavy influence"[19] on all the other great communication routes of the period. In substance, having a recognised and recognisable body would allow the TU to count more on a geo-political level and be able to stand on a par with other telegraph systems in the world.

4.2 Challet-Venel decodes Curchod[20]

In early June 1868 the Head of the Swiss Post Department Challet-Venel wrote to the Federal Council, summarizing the main proposals put

17 It was for example Johann Caspar Bluntschli who in 1853 held that Switzerland was a model "of the fraternization of the Germanic and Romanic character and thus for the future brotherhood of the Germans and the French, on which depends the welfare of Europe". The quotation is from T. Ehs, "An Unwritten History: The Europeanisation of Switzerland" (paper presented at the EUSA 10th Biennial Conference Montreal, Canada, May 17–19, 2007). A modified paper lacking the above quotation is published as T. Ehs, "History in Europeanisation Studies: Lessons from Switzerland," *Contemporary European Studies*, 3/1 (2008).

18 «intermédiaire obligé».

19 «un organe officiel et reconnu» and «influence prépondérant».

20 *Le Département des Postes au Conseil Fédérale*, 3 June 1868, in SFA, E52, 508. Where it is not specified, the quotations in this section are from this document.

forward by other countries and therefore the hottest subjects on the Vienna Conference agenda.

Much space was dedicated to the defence of article 31, which as Curchod had written, had to be protected, though for a different reason. A group of states was in fact intending to present amendments in order to revolutionise the article and fix a lower limit to tariff reductions and therefore free competition. This was a topic that the Swiss government had always had at heart for economic and political reasons\ and why the Postal Department felt the article "had to be stoutly defended"[21].

Challet-Venel then invited the Federal Council to instruct Curchod to defend the 20 word minimum telegram against the 10-word proposal, so that limits were established to curtail Curchod's negotiating possibilities. He also asked for Curchod to be given special powers so that he could sign the final convention like a diplomat.

The little importance the Post Department gave here to the project for creating a special agent is surprising, given the considerable space Curchod had dedicated to it:

> We do not insist on the number 4 which is only a consequence approved by the Federal Council [...] in the sitting of 11 March last. It will be sufficient to remember that following the implementation of the Paris Convention there were very serious irregularities especially in the application of the tariffs and that our suggestions aim at avoiding any repetition, all in order to improve the system of international communications.[22]

There may be many reasons why Challet-Venel did not linger over the creation of the Bureau. As emerges from the passage quoted above, the project was a "natural answer" to a need expressed by the Federal Council, i.e. to improve international telegraphy management. Then Curchod's project was attached to Challet-Venel's communication so that it could

21 «doit être défendu avec insistance».

22 «Nous n'insistons pas sur le chiffre 4 qui n'est qu'une conséquence des propositions approuve par le Conseil fédéral [...] dans sa séance du 11 Mars dernier. Il suffira de rappeler qu'a la suite de la mise en vigueur de la Convention de Paris, il y a eu, surtout dans l'application des tarifs, des très graves irrégularités, et que nos propositions ont pour but d'en éviter le retour, tout en améliorante le système des communications de service internationales».

speak for itself. Finally, the Post Department had such a high esteem of Curchod that its approval of his well-argued proposal was implicit, a surmise that appears to be backed by the fact that Curchod obtained full powers as delegate to the conference. The instructions he received from the Federal Council were limited to a summary of Challet-Venel's suggestions and before Curchod had drawn up his project, he was authorized "not to hold himself subject to the limits set by the present instructions"[23].

What may seem surprising is the fact that the Federal Council did not emphasize the strategic importance for the Swiss government to control the Bureau, as instead Curchod gave to understand in the final part of his project. Challet-Venel may not have emphasized strategy because quite simply the idea that Switzerland was to run the office was not yet mature, as is evidenced by the fact that Challet-Venel always wrote of the "Administration of the State in whose capital the last Conference took place"[24] to indicate the authority the new body would be placed under.

4.3 The anti-dumping regulation

In Vienna Curchod promoted two norms that were to be crucial for the running of the international service: an anti-dumping tariff regulation in article 31 and the constitution of a permanent body of the TU, named *Bureau International des Administrations Télégraphique* (regulated by article 55 of the new convention).

Article 31 of the Paris Convention, which both the Federal Council and the Post Department wanted to keep, was the subject of long discussion because various states desired to fix limits on the freedom of single countries to lower both terminal and transit tariffs. The delegates

23 «de sortir des limites imposées par les présentes instructions». *Instructions pour les Conférences de Vienne*, 5 Juin 1868, in AF, E52, 509.

24 *«Administration de l'Etat dans la Capitale duquel la dernière conférence a eu lieu».*

from Northern Germany, Persia, Belgium, Italy and Holland, submitted an amendment with a very explicit aim:

> When it is a matter of correspondence borrowing the lines of more than two states, the terminal and transit charges can only be modified by the contracting and member states.[25]

As the Belgian delegate Vinchent wrote, this change was proposed with the main purpose of "making impossible abnormal competition caused by simply lowering tariffs. [...] At the levels tariff rates have gone down to, private interest has nothing to gain from these immoderate reductions, which compromise investment in depriving them of all profit"[26]. It was a real and proper anti-dumping regulation.

As recorded in Chapter 3, the Paris Conference had established the criterion for telegraph operators to follow. Where there were different routes on offer, which often happened given the extreme ramification of the international network, the cheapest route had to be chosen. The rule had left the various countries the freedom to reduce international tariffs and therefore be at an advantage when routes were being chosen. It is what the Swiss had done in introducing the lowest transit fee, so that it was good value and then compulsory for telegrams to be routed through Switzerland. Over time, however, the compulsory cheapest route could well damage user interests in that telegrams travelling at rock bottom prices could be sent via routes that gave no guarantees over time or accuracy of the message sent[27].

It was to protect the clientele, therefore, that at Vienna the anti-dumping amendment was promoted to avoid excessive competition among administrations via indiscriminate tariff cutting. As declared

25 «Lorsqu'il s'agit de correspondances qui empruntent les lignes de plus de deux Etats, les taxes terminales et des transit ne peuvent être modifiées que du consentement unanime des Etats contractants et adhérents». *Documents de la conférence télégraphique internationale de Vienne*, p. 305.

26 «rendre impossibles des concurrences anormales, réalisé par le seul abaissement des taxes. [...] Au taux auquel sont descendus les tarifs, l'intérêt privé n'a rien à gagner à des réductions immodérées, qui compromettent les exploitations en les privant de tout bénéfice». *Ibidem*, p. 307.

27 Fari, *Una penisola in comunicazione*, 487.

again during the following Rome Conference, the aim was essentially to improve as far as possible the quality of the service. If a new line opened in competition with an already existing one and a lower tariff was offered, there would be a dual effect. Users would go for the cheaper way and the older line would be abandoned with a serious waste of the public money used to build it. Yet, it was in the public interest to be able to access various routes and then opt for speed or cheapness[28].

In spite of the promoters' good intention, the "anti-dumping" amendment caused a fracture among the delegates in Vienna, with two lines of thought coming to the surface. The first defended article 31 as it had been drawn up in Paris while the second proposed curbing the freedom to lower tariffs on new telegraph lines. The first position, held by the states with solid liberal tradition, especially France and Switzerland, is well summarised in the words of the Danish delegate Nielsen:

> These restrictions to the freedom of states are extremely serious; they would make too radical a change to one of the essential principles of the Convention, by which the Conference alone is mandated to carry out revisions and has to right to use them.[29]

The second line of thought, held both by conservative countries like Turkey and Austria and industrialized ones like Belgium and Holland, is well reflected in the words of the Turkish delegate, who insisted "on the inconveniences of the present system which authorizes the most exaggerated competition"[30].

The division over the amendment to article 31 reflected the two conflicting attitudes at the Vienna Conference. On one side, Switzerland and partly France conceived of the telegraph administration as a state

28 *Documents de la conférence télégraphique internationale de Rome*, Publiés par le Bureau international des administrations télégraphiques, Imprimerie Rieder & Simmen, Berne 1872, pp. 386–388.

29 «Ces restrictions à la liberté des Etats sont extrêmement graves; elles modifieraient trop radicalement un des principes essentiels de la Convention, pour que la Conférence, qui est chargée seulement d'un travail de révision, ait le droit de les adopter». *Documents de la conférence télégraphique internationale de Vienne*, p. 314.

30 «sur les inconvénients du système actuel qui autorise les concurrences les plus exagérées». *Ibidem*, p. 315.

company and as such subject to the laws of free competition. On the other, Belgium, Italy, Turkey and Holland had a more interventionist mind cast and treated the telegraph service as a ramification of state organization. Thus, the telegraph administration was likened to all other public administrations and had to be more or less subject to the interests of the government and state. In this case, free competition might well take away clients from many administrations, which would suffer a drastic loss of income from international telegrams. This was quite a serious matter, given that many administrations including the Swiss one managed to post profits only because of income from international traffic.

Contrast between the two concepts of telegraph service was so strong over article 31 that the chairman decided to put the question off until a later sitting. Nevertheless, in order to reach an agreement, which was impossible during the plenary, it was decided to entrust the drawing up of a new article 31 reconciling the two positions to a commission revising the convention.

Switzerland, which had made the progressive reduction of tariffs a mainstay of its management of international telegraphy, sided against the amendment as proposed because it risked endangering its policy and privileges built up over the past. Curchod illustrated the reasons why he could not back the amendment, though leaving a final aperture:

> The instructions of M. Curchod [...] would oblige him to defend the system now in force [...]. M. Curchod however shares the general opinion that it is by the nature of the service and not exaggerated reductions in charges that the states must compete against one another and while keeping to the principle of freedom, he would refuse to introduce a restriction in thus sense.[31]

Curchod upheld that if this amendment made Switzerland lose income from transit telegrams, he would be unable to accept it, partly because

31 «Les instructions de M. Curchod [...] l'obligerait à défendre le système actuellement en vigueur [...]. M. Curchod partage toutefois l'avis général que c'est par la nature du service et non par des réductions de taxes exagérées que les Etats doivent lutter entre eux, et, tout en maintenant le principe de liberté, il ne refuserait pas d'y introduire une restriction dans ce sens». *Ibidem*, p. 308.

of the instructions received from the Federal Council. If the main aim was for the general good, i.e. improve the public service, he would not oppose the modification.

Even though there are no extant minutes of the commission debates, three versions of the amendment were probably discussed. Curchod's proposal, which attempted to reconcile the freedom of the single states with the introduction of an anti-dumping law, came out victorious at the end:

> The charges contained in these tables can, always and in any period, be reduced by common agreement between one or other of the governments concerned, nevertheless these reductions must have for aim and effect not to create competition in charges among existing lines but to open to the public with equal charges as many lines as possible.[32]

In the most significant part, it was established that telegraph tariffs could be reduced at any moment on mutual agreement between the countries involved, and therefore without the general consent required in the first version of the amendment. However, the last sentence established that these reductions could only be applied in the case of a new line opened on an entirely new route. The new amendment was proof of Curchod's diplomatic skill, fruit of his great capacity to mediate between liberal principles and the specific interests of the member states. His success testifies two factors. First, it points to his capacity to go beyond received instructions. Even though the Federal Council had invited him to defend article 31 as it stood with the freedom of all states to change telegraph tariffs, Curchod perceived the general desire for modification and did not hold back, but took steps in favour of Switzerland. In effect, this new formulation was decidedly pro Switzerland because it froze the present situation. At the most, tariffs could be reduced to the same value as on existing potentially competitive routes. This meant that opening a

32 «Les taxes inscrites dans ces tableaux pourront, toujours et à toute époque, être réduites d'un commun accord entre tel ou tel des Gouvernements intéressés, toutefois ces réductions devront avoir pour but et pour effet, non point de créer une concurrence de taxes, entre les voies existantes, mais bien d'ouvrir au public à taxes égales autant de voies que possible». *Ibidem,* p. 411.

new telegraph line would not set off competition, but only increase the possibility of choice without damaging the networks and investments already made. As Curchod would confirm in the report on Vienna he sent to the Federal Council, the new formulation of the article "furthermore answers all the points of the ideas that have always prevailed in the efforts our administration has made to attract as much transit traffic as possible to our lines"[33]. Thus, Switzerland had protected itself from possible tariff modifications and even when faced by reductions made by other states, its transit tariffs would still be extremely economical and able to attract much traffic through its network.

4.4 The debate over the Bureau

Curchod was also the leading figure in the debate over article 55, which regulated the relations between the TU and member states. The discussion was particularly thorough given that the article determined how in between conference periods member administrations would exchange information both about network changes (new lines, office hours, cable laying) and tariffs reductions as well as proposals for changing or interpreting regulations/conventions[34]. As seen at the Paris Conference, France had proposed setting up a special commission of members representing national telegraph administrations in the TU. It would operate in the city where the latest conference had been held and which would meet on special occasions between one conference and the next. Meetings would be called whenever serious doubts arose and non-participating states would have to abide by the commission's decisions. The proposal had found no favour in Paris, but the French

33 «répond d'ailleurs en tous points aux idées qui ont toujours prévalu dans les efforts que notre administration à faits pour attirer par ses lignes le plus de transit possible». *Lettre écrite par Curchod adressée à Challet-Venel*, 22 Août 1868, in SFA, E52, 509.

34 *Documents de la conférence télégraphique internationale de Vienne*, pp. 379–398.

delegate presented it in more or less the same form in Vienna with the following motivations

> Differences of opinion could arise over the exact meaning of some of the Convention's dispositions. Now experience shows that in similar cases agreement is more difficult to reach by correspondence and that debate in conferences leads to quicker and more decisive results. Therefore, if one of these difficulties arises, it would be of benefit to be able to have it resolved by a commission. [...] Its usefulness is beyond doubt.[35]

The French delegate Jagerschmidt recognised that the system of reciprocal communications between nations established at Paris was cumbersome and betrayed signs of evident inefficiency. In fact, any network or tariff changes had to be communicated to all member countries, which resulted in an unnecessary flow of information, little used by the telegraph managers because its meaning was difficult to grasp. Inefficiency at this level in the TU structure provided the springboard for the Swiss to present their amendment.

In fact it was just at this moment that Curchod took the word, identifying "two gaps [...] to examine separately"[36] in the Paris Convention. One gap concerned the convention as a whole, in that there was no way of making changes between one conference and another and this would be filled by the commission proposed by the French. The second gap concerned the regulation and regular running of international telegraphy. Curchod felt that, in this case, a body needed to be created to deal with administrative matters. His proposal, therefore, was complementary to the French one, not an alternative.

Curchod read out from notes he had prepared a description of his idea of a "telegraph agency". He began by underlining that "the Paris

35 «Des divergences d'opinion peuvent se produire sur le sens exact de quelques dispositions de la Convention. Or, l'expérience démontre qu'en pareil cas l'accord s'établit difficilement par correspondance, et que les discussions en conférence conduisent à des résultats plus prompts et plus décisifs. Si donc une de ces difficultés se présentait, il y aurait intérêt à pouvoir la faire résoudre par une Commission. [...] Son utilité paraît incontestable». *Ibidem*, pp. 379–380.

36 «deux lacunes [...] d'examiner séparément». *Ibidem*.

Convention and the Service Regulations which complete it must be applied in the telegraph networks of Europe, Asia and Africa by twenty-six administrations"[37]. This was the reason why, he held, there were differences of opinion between the twenty-six administrations. In this way, he took up the French project and emphasized the need for a body to govern the relations between administrations, highlighting particularly that "the French proposition failed, not because the Conference disapproved of it but because the organisation of the commission was not sufficiently resolute"[38]. This was intended to emphasize that Swiss telegraphy was of a higher level than the French one

> Switzerland has taken up again [the 1865 French one] carrying out a modification so as to prevent the objections that it had caused. Under the management of the telegraph administration of the State where the most recent conference has taken place, there would function for everything concerning the service and information about the fulfilment of the Convention, a special agent, named by the Conference and financed by all the contracting states. This agent, to whom the title of general secretary of the Conferences would be given, would replace for everyday and run of the mill business the Commission – Commission which could in any way convene, but be convened in the unlikely case of a serious difficulty.[39]

Then, linking up with what Jagerschimdt had said, Curchod explained that the moment had by then arrived to create a standing body since

37 «la convention de Paris et le Règlement de service qui la complète doivent être appliqués dans les réseaux télégraphiques de l'Europe, de l'Asie et de l'Afrique, par vingt-six Administrations». *Ibidem,* p. 381.

38 «la proposition de la France a échoué, non pas que la Conférence en désapprouvât le but, mais parce que l'organisation de la commission n'était pas suffisamment déterminée». *Ibidem*, p. 382.

39 «La Suisse a repris cette [the 1865 French one] en y apportant une modification de nature à prévenir les objections qu'elle avait soulevées. Sous la direction de l'Administration télégraphique de l'Etat où la dernière Conférence a eu lieu, fonctionnerait, pour tout ce qui concerne le service des communications et renseignements relatifs à l'exécution de la Convention, un agent spécial, nommé par la Conférence et à la solde commune de tous les Etats contractants. Cet agent, auquel on donnerait par exemple le titre de Secrétaire–général des Conférences, remplacerait pour les affaires courantes et de peu importance la Commission dont il vient d'être question, Commission que rien n'empêcherait cependant de constituer aussi, mais pour n'être appelée à siéger que dans le cas peu probable où une difficulté sérieuse viendrait à se présenter». *Ibidem.*

"The difficulties arising among the Administrations in applying the Paris Convention derive essentially if not uniquely misunderstandings, insufficient or ill understood information"[40]. Curchod went on to cite cases of misunderstandings caused by the simple need to communicate tariff changes or office hours to all member states and so in his sober, liberal style he then proposed a practical solution – streamlining:

> Things would work differently with the central agency Switzerland proposes. Each Administration would simply contact by letter as often as necessary, with the communications in question; and periodically under the same known form, the agency would send around a circular furnishing all the information it has received.[41]

The agency would deal with the international nomenclature and statistics, which called for a single, centralized programming mediating relations between TU members and private companies[42]. Meaningfully Curchod went on to list the powers and limits that the "special agent" would have, which he then found himself working with a few weeks later in the role of Bureau Head: "he will be under the orders of the administration in charge and will work with it to study all the questions touching common interests. He will attend the conferences in an advisory capacity"[43].

Switzerland's lengthy amendment, identical to the project Curchod had shown Challet-Venel in the months running up to the conference, provided for the creation of a Conference-nominated special agent, salaried by all member states, under the "administration in charge" and

40 «Les difficultés qui se sont présentées entre Administrations dans l'application de la Convention de Paris, sont résulté essentiellement, si ce n'est uniquement, de malentendus, de renseignements insuffisants ou mal compris». *Ibidem,* pp. 382–383.

41 «Les choses se passeraient différemment avec l'agence centrale que propose la Suisse. Chaque Administration lui adresserait par simple lettre, aussi souvent que cela lui conviendrait, les communications dont il s'agit; et périodiquement, sous une forme connue et toujours la même, l'agence expédierait une circulaire fournissant tous les renseignements qu'elle aurait reçus». *Ibidem*, p. 384

42 *Ibidem*, p. 385.

43 «il serait sous les ordres de l'Administration directrice et fonctionnerait auprès d'elle pour l'étude de toutes les questions touchant aux intérêts communs. Il assisterait aux Conférences avec voix consultative». *Ibidem*.

mandated to deal with everyday business. The main function was to avoid misunderstandings, sort out the more obscure administrative issues and "even go further in the matter of tariffs"[44].

The most substantial support for Curchod's project came from Vinchent, whose series of considerations helped the project to pass without any difficulty. First, he recalled that Curchod's project was not a complete novelty[45], since examples of similar institutions already existed in railways.

> The agent-general of the Rhenish, Belgian and Northern French Company railways, an agent living in Cologne, supplies each of the three administrations with the information they need, deals with matters common to their networks and finds solutions: in the case of disagreement he gives his advice, but never takes any decision. This role is similar to the one to be filled by the new telegraph agent.[46]

Secondly, Vinchent traced out the profile of the ideal candidate which fully corresponded to a description of Curchod himself:

> To be appropriately carried out, a similar mission would call for a man combining experience of the telegraph services and business with recognized merit and a knowledge of several languages. But he would be difficult to find if he is obliged to change residence every three years and move from capital to capital. The essential condition for a good choice is a stable residence.[47]

---

44 «même aller plus loin en fait de tarifs». *Ibidem,* p. 384.

45 Today, Curchod's profile is considered the prototype for the modern permanent secretariat as established by an international treaty, see A. Peters and S. Peter, 'International Organizations: Between Technocracy and Democracy', in *The Oxford handbook of the history of international law*, ed. B. Fassbender and A. Peters (Oxford: Oxford University Press, 2012), 175–176.

46 «l'Agent général des chemins de fer Rhénans, de l'Etat Belge et de la Compagnie française du Nord, (est) un agent résidant à Cologne fournit à chacune de ces trois Administrations les renseignements dont elle besoin, s'occupe des affaires communes à leurs réseaux, et en prépare les solutions; en cas de désaccord, il donne son avis, mais jamais prendre aucune décision. C'est un rôle analogue que devrait remplir le nouvel agent télégraphique». *Documents de la conférence télégraphique internationale de Vienne,* p. 388.

47 «Une pareille mission exigerait, pour être convenablement remplie, un homme joignant l'expérience du service télégraphiques et des affaires à un mérite reconnu, connaissant même plusieurs langues. Mais on le trouverait difficilement si

In order to find such a figure, however, Vinchent felt a "permanent seat"[48] was necessary, so the agent was not faced with a change of residence every three years and a move to the capital of the next country hosting the Conference. Though this point was close to Curchod's heart, he had not brought it up, presumably for political expediency[49].

The North German Confederation delegate, de Chauvin, who also judged the agent as profiled by Curchod indispensable, added that the appointee would have to deal with editing a telegraph journal containing telegraph statistics (the future *Journal Télégraphique*) and meaningfully suggested collocating the seat of the agency "in a *neutral city*"[50]. Here too Switzerland, with its status of perpetual neutrality, seemed of course to fit in perfectly with the profile described by de Chauvin. The French delegate Jagerschmidt expressed his conviction that international tariff statistics needed to be organized, as well as a journal publishing news on telegraph developments for all the administrations.

He was less convinced of the need to set up the figure of a general agent, and when Curchod detailed his proposal, there was a moment of friction in a debate which otherwise went much in favour of the Swiss suggestion. Partly thanks to the support of the other delegates, Curchod came out into the open and declared that the:

---

l'obligation de changer de résidence tous les trois ans et de se transporter de capitale en capitale lui était imposée. La fixité de la résidence est la condition essentielle d'un bon choix». *Ibidem*, [*our italics*].

48 «fixité de la résidence». *Ibidem*.

49 Like Vinchent's unconditioned backing, this episode could give rise to the surmise that a kind of iron pact had been created between Belgium and Switzerland to get Curchod's project approved, and that anyway Vinchent was well acquainted with and approved of Curchod's aims. There are no documents offering any proof, however.

50 «de la placer dans un *ville neutre*». *Ibidem*, p. 389, [*our italics*].

JOURNAL
TÉLÉGRAPHIQUE
LE BUREAU INTERNATIONAL
N° 1.

Figure 12: The first issue of the *Journal Télégraphique*, edited and published by the Bureau starting from 1869 and today still active.

Source: <http://gallica.bnf.fr>.

> General Secretary exercising his functions beside the chief of the head office and under his authority and responsibility [...] would move in *a wider sphere* outside all control. The work on statistics would be the secondary part of his task; he would be above all *advisor to the Administration, the regulator of the international service.*[51]

This job description of the general secretary disconcerted Jagerschmidt, who openly accused Curchod and Switzerland of being illiberal:

51 «Secrétaire-général exerçant ses fonctions à côté, sous l'autorité et sous la responsabilité du chef de l'office Directeur [...] se mouvrait dans une *sphère plus large*, en dehors de tout contrôle. Les travaux de statistique seraient la partie secondaire de sa tâche; il serait surtout le *conseiller des Administrations*, le *régulateur du service international*». *Ibidem*, p. 390, [*our italics*].

M. Jagerschmidt rejects such a situation, which he feels is *incompatible with the dignity and liberty of telegraph administrations.* The general agent, more independent than the different offices and unanimity of the Conference necessary to appoint him, will be at the same level to take over its functions, and will not take long to become an obstacle for the international service instead of a force working for it.[52]

He identified at least three negative features in the Swiss proposal. Firstly, such an extension of the secretary general's functions could lead to a curtailing of the decisional power of the single states. The figure could consequently acquire an excessive power that would be difficult to keep under control. Furthermore, and more subtly, the French delegate feared that a telegraph agency as conceived by Curchod could invalidate his own project of an itinerant special commission. Indeed, as the debate continued and it appeared clear that the idea of the special commission would be approved too, he stopped making objections.

Answering accusations of a lack of liberalism in Curchod's approach, Vinchent replied at once and took advantage of the occasion to note the need to entrust the appointment to an eminent person of high moral stature:

M. Vinchent answers that as the agent general has no decision to take, no way of imposing his opinions, the freedom of the Administrations would remain intact. He would exercise *a purely moral influence, never greater than a recommendation of advantage, coming from the authority of his position, experience and capacity.* It is for that that M. Vinchent would like to give the new institution a genuine importance, create it independent and enhance it by choosing a man esteemed for his character and personal value.[53]

---

52 «M. Jagerschmidt repousse une pareille situation, qui lui paraît *incompatible avec la dignité et la liberté* des Administrations télégraphiques. L'agent général, d'autant plus indépendant des divers offices que l'unanimité de la Conférence, nécessaire pour le nommer, le serait également pour lui retirer ses fonctions, ne tarderait pas à devenir un obstacle pour le service international au lieu d'être une force pour lui». *Ibidem*, [*our italics*].

53 «M. Vinchent répond que, l'Agent général n'ayant aucune décision à prendre, aucun moyen d'imposer ses avis, la liberté des Administrations demeurerait entière. Il exercerait *une influence purement morale, mais d'autant plus grande qu'il se recommanderait d'avantage, par l'autorité de sa position, son expérience et sa*

Serpos Effendi, the Turkish delegate, suggested the administration hosting the last conference should deal with the functions Curchod and Vincent attributed to the general agent. At this point, the word was taken by Brunner, the head of the telegraph department of Austria, the host country responsible for appointing the Secretary General and setting up the International Bureau. He declared that he could not see "anything against this officer being nominated by the Conference, providing he is the secretary, the agent of the Director's Office"[54]. This apparent lack of interest in the nomination of the International Bureau Head, which was *de facto* a renunciation to a right, was praised by the French delegate, who however invited Brunner to change his mind because he held that "from the viewpoint of the hierarchy and the interests of the service [...] this nomination is not to be made by the Conference but by the Office in charge"[55]. Jagerschmidt probably wanted the Bureau head to be appointed by the country hosting the conference and not by the conference itself, in order to lower the Bureau's prestige. If the plenary assembly elected the head, its vote would legitimate the power of the head and the Bureau.

This impasse was removed by the proposal of the delegate from Baden, Zimmer, who held the office "had to be attached to *one* Administration, so that it finds itself, so to speak, surrounded by everyday current affairs"[56]. Staring, the Dutch delegate, summed up in an amendment that "a Conference-designated telegraph administration" would have in its service the International Bureau and, only in this way, would it have "the stability and all the advantages that had been called for in

capacité. C'est pour cela que M. Vinchent voudrait donner une importance véritable à l'institution nouvelle, la constituer indépendante et la rehausser par le choix d'un homme honoré à cause de son caractère et de sa valeur personnelle». *Ibidem*, p. 391, [*our italics*].

54 «aucun inconvénient à ce que ce fonctionnaire soit nommé par la Conférence, à la condition qu'il soit le secrétaire, l'organe de l'office Directeur». *Ibidem*, p. 392.

55 «au point de vue de la hiérarchie et de l'intérêt du service, [...] cette nomination ne soit pas faite par la Conférence, mais bien par l'office Directeur». *Ibidem*, p. 393.

56 «doit être rattache à *une* Administration, afin de se trouver, pur ainsi dire, au milieu des affaires courants de chaque jour». *Ibidem*.

the creation of a Secretary-general"[57], as had happened for the Union of the German Railways.

At the end of the debate, the nature of the new body was clarified and summed up in the new article 61:

> A conference-appointed Telegraph Administration will take the measures to facilitate the implementation of the Convention. To do this, it will organize under the name of "International Bureau of Telegraph Administrations" a special service which will operate under its orders [...]. It will gather all kinds of information about international telegraphy, draw up tariffs, prepare general figures, go ahead with research of common utility it perceives, and edit a French language telegraph paper.[58]

Although the name of the administration entrusted with hosting the Bureau had not yet been made, all the elements emerging during the debate pointed to the natural candidature of Switzerland. It was in fact the only neutral country having in its service a well-esteemed and polyglot technician with a long experience in the telegraph service like Curchod. With very little discussion at all, the Vienna Conference unanimously designated the Swiss Post Office as the administration the International Bureau would depend on. The plenary also "declared its formal wish to see M. Curchod placed at the head of the Bureau"[59]. The latter first heard the Federal Council and then formally accepted the appointment on 21 July 1868.

The Vienna Conference concluded with a triumph for Curchod, well summarised by the Prussian delegate who wanted to emphasize the

57 «une Administration télégraphique désignée par la Conférence» and «la stabilité et tous le avantages qu'on avait demandés à l'institution d'un Secrétaire-général». Ibidem, p. 394.

58 «Une Administration télégraphique, désignée par la Conférence, prendra les mesures de nature à faciliter, dans un intérêt commun, l'exécution et l'application de la Convention. A cet effet elle organisera, sous le titre de "Bureau international des Administrations télégraphiques", un service spécial qui fonctionnera sous sa direction [...] Il centralisera les renseignements de toute nature relatifs à la télégraphie internationale, rédigera le tarif, dressera une statique générale, procédera aux études d'utilité commune dont il serait saisi, et rédigera un journal télégraphique en langue française». *Ibidem,* pp. 433–434.

59 «émis le vœu formel de voir M. Curchod placé à la tête de ce bureau». *Ibidem,* p. 454.

prestige of the new institution by having the name of the new director and his salary explicitly voted by the Conference:

> He would like, on the contrary to facilitate exercising his right and in front of certain scruples that the Federal Council will certainly sustain in giving him the assurance that if his free choice fell on M. Curchod, the Conference which has been able to appreciate all he has of knowledge and experience, would have a future pledge for the creation of the International Bureau, which he considers the most important result of the Conference's works; in letting him know, besides, that the services of the Head of this Bureau do not seem to him to be too highly remunerated with an annual salary of twelve thousand francs.[60]

On 25 July 1868 Brunner, Head of Telegraphs of the country hosting the conference and ex-head of Swiss Telegraphs, sent the President of the Swiss Confederation, Jakob Dubs, a letter in which he explained some of the reasons why the delegates at Vienna had chosen Switzerland to run the Bureau. While some reasons had already emerged in the course of the conference debate (e.g. Switzerland's neutrality[61]), Brunner gave two others which are not mentioned in the minutes. The conference probably held that, Switzerland's "geographical position"[62] made it the ideal candidate for running an efficient service for the general good. Secondly, Brunner thought the choice had fallen on Switzerland because the delegates wanted to "pay homage to the precision of the telegraph service so well exemplified by Switzerland"[63]. In other words, a last reason that took Switzerland to the head of the Bureau was that

60 «Il voudrait, au contraire, faciliter l'exercice de son droit et aller au-devant de certains scrupules que le Conseil fédéral éprouvera certainement, en lui donnant l'assurance que, si son libre choix tombait sur M. Curchod, la Conférence, qui a pu apprécier tout ce qu'il ya chez lui, de savoir et d'expérience, y verrait un gage d'avenir pour l'institution du Bureau International, qu'il considère comme le résultat le plus important des travaux de la Conférence; en lui faisant savoir, en outre, que les services de Directeur de ce bureau ne lui paraîtraient pas trop largement rémunérés par un traitement annuel de douze mille francs». *Ibidem*, p. 456.

61 "political position" («position politique»), *Lettre écrite par Brunner adressée à le président de la Confédération Suisse Dubs*, 25 Juillet 1868, in SFA, E52, 510.

62 «situation géographique». *Ibidem*.

63 «rendre hommage à l'exactitude du service télégraphique dont la Suisse donne un si bel exemple». *Lettre écrite par Brunner adressée à Dubs*.

the international community recognised the efficient management of its domestic telegraph service.

Political neutrality, geographical centrality, multilingualism, Curchod's recognized skills as head of telegraphs, and more generally the technical efficiency of the country. These were the most explicit reasons giving Switzerland the management of the international office that would control the European and then world network.

Figure 13: Louis Curchod: Head of Swiss Telegraphs and first secretary general of the International Bureau of Telegraph Administrations.

Source: ITU Archives.

4.5 Swiss reactions

In August 1868 about a month after the end of the Vienna Conference, Curchod sent the Post Department a document in which he summed up the main results obtained. Besides the already mentioned description of the changes in Switzerland's favour of article 31 and a long disquisition on tariff changes, Curchod concentrated on the discussions over the founding of the Bureau and so revealed some of the backstage activity, which did not emerge in the debates.

First, he informed the government that the French project for a special commission had been approved. He felt however that it was a minor project, destined to failure because of "important differences on the interpretation of the Convention are unlikely, of the kind that the Commission would hardly have time to convene and [...] for ordinary business and everyday affairs it would be of no use"[64].

For everyday running Curchod had managed to get the approval for an international office, a project he personally drew up but "meeting the Swiss Administration's proposition approved by the Federal Council in the sitting of 11 March 1868"[65]. The Swiss government had in fact called for the creation of an agency headed by a general secretary nominated by the Conference and under the direction of the "administration in charge"[66], in order to streamline the everyday affairs of the Union. Curchod then detailed how the delegates had welcomed this project enthusiastically:

> Very soon after my arrival in Vienna, I had the occasion to ascertain the idea thus put forward was circulating. Some very influential members of the Conference were declaring they would support it with all their power and that they would consider its realization the most important result our works could lead to.[67]

---

64 «des dissentiments importants sur l'interprétation de la Convention sont peu probables, de telle sorte que la Commission n'a guère de chances de se réunir et [...] pour les cas ordinaires et pour les affaires courantes elle n'aura aucune utilité». *Lettre écrite par Curchod adressée à Challet-Venel*, 22 August 1868.

65 «en satisfaisant à la proposition de l'Administration Suisse, approuvé par le Conseil fédéral dans sa séance du 11 Mars 1868». *Ibidem.*

66 «administration en charge». *Ibidem.*

67 «Bientôt âpres mon arrivée a Vienne, j'eus l'occasion de constater que l'idée ainsi mise en avant trouvait de l'écho. Des membres très influentes de la Conférence déclaraient qu'ils l'appuyéraient de tout leur pouvoir et qu'ils considéraient sa réalisation comme *le résultat le plus important* auquel nos travaux puissent aboutir». *Ibidem.* De Chauvin expressed himself in a similar way in the conference's closing

He recalled how the basic principles of the Swiss proposal had remained the same during the debate and only the form had changed, in order to "satisfy the wish of some delegates"[68]. Instead of a central agency and a conference-nominated general secretary subjected to the orders of the administration hosting the plenary, approval had been given to a Bureau with its own head named by an administration "that would be mandated to organize this service"[69]. It seems from these words that substantially Switzerland found itself almost involuntarily in charge of the management of the Bureau and that the Federal Council had not thought to obtain it, but only to propose the centralization of the everyday running of international traffic simply to remove it from any particular interests. During the Conference, however, Curchod found himself in the position of obtaining presidency, seat and administrative management of the Bureau at the dependence of the Swiss Post Department and grasped it at once. Whatever, on 20 July 1868 (the day after Switzerland and Curchod were unanimously voted the ideal candidates for hosting and running the Bureau) the Federal Council "accepted the mission"[70].

Even the federal councillor heading the Post Department, Challet-Venel, did not seem to wholly understand the political/strategic relevance of the Bureau. In a letter to the Council, he affirmed that "the main change" of the Conference was the creation of the Bureau, but at the same time, he maintained that the Bureau was no more than "an administrative matter"[71].

More than Curchod himself, more than the Post Department and more than the Federal Council, the importance of having obtained

debate, calling the founding of the International Bureau "the most important result of the Conference" («le result le plus important de la Conference»). *Documents de la conférence télégraphique internationale de Vienne*, p. 456.

68 «satisfaire au désir de quelques déléguées». *Lettre écrite par Curchod adressée à Challet-Venel*, 22 Aout 1868.

69 «qu'on chargerait de bien vouloir organiser ce service». *Ibidem*.

70 «accepté cette mission». *Ibidem*.

71 «la modification la plus essentielle» and «une affaire purement administrative». *Département fédéral des Postes au Conseil fédéral*, 3 Septembre 1868, in SFA, E52, 508.

control over the Bureau was understood by the National Council Commission charged with evaluating the work of the Federal Council and Federal Tribunal. This body, an expression of the main Swiss political forces and with therefore a less pragmatic role than the Federal Council, recalled that control over the Bureau could be collocated in a wider strategy seeing Switzerland in the front line for peace in Europe. This would lead to an increasing importance on the European political stage:

> The establishment of the International Telegraph Bureau in our country proving Switzerland enjoys a greater and greater prestige among the members of the European family.[72]

The Bureau was, therefore, one example of how during the nineteenth century, Switzerland learnt how to acquire an ever increasing importance and recognition within Europe. This was not only because it lay geographically right in the centre of the continent and therefore the international telegraph network, but really because it desired to carve itself out an increasingly visible international role. It was what the historian Madeleine Herren defines "governmental internationalism", i.e. a specific foreign policy Switzerland put in act in the second half of the nineteenth century in order to increase its own international importance and power, though being overall a small peripheral state in comparison to the great powers of the period. The main way Switzerland applied governmental internationalism and acquired prestige and international recognition was to specialise in the management of supranational bodies. The TU was in fact only the first of the many international organizations which in the second half of the nineteenth century and throughout the twentieth century were assigned to Switzerland, which saw in these bodies a chance to acquire greater visibility and prestige on an international level[73].

---

72 «l'établissement du Bureau télégraphique international dans notre pays prouvent que la Suisse jouit d'une considération toujours plus grande parmi les membres de la famille européenne». FF, 2/23, 26 Juin 1869, p. 229.

73 M. Herren, "Governmental Internationalism and the Beginning of a New World Order in the Late XIX Century," in *The Mechanics of Internationalism: Culture, Society, and Politics from the 1840s to the First World War*, ed. M. H. Geyer and J. Paulmann (Oxford: Oxford University Press, 2001), pp. 132–135.

4.6 Organization and ambiguity

Though the tasks of the new international office were clearly listed in Curchod's Vienna-approved project, it was still left to establish how it would be organized institutionally. The Vienna Conference had already expressed its opinion and in the last sitting of 21 July furnished some guidelines to the Federal Council. It suggested the Bureau's Head should depend not on the Head of Swiss Telegraphs, but directly on the Head of Post Department. It also recommended an annual salary of 12,000 francs[74].

On 30 September 1869, the Head of Post Department Naeff, back in office after the brief Challet-Venel interlude, declared he was in complete agreement with the indications given in Vienna over the position of the new Bureau head. He would be "completely independent from the Federal Director of Telegraphs (Head of Telegraphs) and that he would be placed directly under the orders of the Federal Councillor in charge of the Higher Direction of Telegraphs"[75].

Within the particular Swiss institutional set up, however, the Federal Council played a role like the Directory governing France post 1789[76]. Though the seven ministers each ran a department, they were called to exercise their executive power in a collegial way, so that the various departments were equally important and the councillors responsible for them all had the same power within the government. It followed that the International Bureau would depend not only from the Post Master, but also from the Federal Council all together, which *de facto* could choose autonomously the direction and complex of strategies to adopt.

74 *Documents de la conférence télégraphique internationale de Vienne*, p. 456.

75 «dans une position complètement indépendant du directeur fédéral des télégraphes et qu'il serait placé immédiatement sous les ordres du Conseiller fédéral chargé de la direction supérieure des affaires télégraphiques». *Département fédéral des Postes au Conseil fédéral,* 30 Septembre 1868, in SFA, E52, 510.

76 See for example H. Kriesi, *Le système politique suisse* (Paris: Economica, 1998).

A second feature Naeff concentrated on was the Bureau's budget. Although the Vienna Conference had indicated in the sum of 40,000 francs the Bureau's first annual budget, he supplied the documentation to prove that outgoings could be reduced to 31,000 francs, close to the sum quoted by Curchod in his original spring 1868 project. Curchod's salary was also affected by the suggested cuts. Instead of the 12,000 francs recommended by the Conference, he would receive 8,000 "only in view of observing the hierarchy of treatment and so that the honorarium of the Bureau head would not be higher than that of the Head of the Post Department"[77]. Naeff, who would be supervising the Bureau Head's work, was preventing Curchod from earning more than he did[78].

The overall structure of the Bureau would be very streamlined. Naeff's plan provided for the head and only three other permanent employees: a secretary (responsible for compiling the tariffs and statistics as well as editing the journal), a technician specialized in telegraph and electric matters and a copyist. Other copyists and planners would be taken on later, according to need.

In a letter that Naeff sent to Curchod on 14 October 1868, the powers and particularly the limits of the Bureau Head were described in detail. As Federal Councillor responsible for the Post Department, Naeff reserved for himself the power to nominate staff for the Bureau, also on the advice of the Bureau Head, who could present his own list of candidates[79]. The memorandum was also extremely clear about the relations between the Bureau and the Head of Swiss Telegraphs. The latter

77 «uniquement en vue de conserver l'ordre hiérarchique des traitements et afin que les honoraires du Directeur du bureau International ne surpassent pas ceux du chef même du Département des postes». *Département fédéral des Postes au Conseil fédéral,* 30 Septembre 1868.

78 The close attention paid to any possible limitation of public spending was in line with the concept of the body politic in nineteenth century Switzerland. The Democratic Radical Party, which dominated Swiss political life from 1848 up to WWI, counted on strengthening a central power capable of guaranteeing secular institutions and the formal equality of the citizens, as well as inheriting from classical liberalism the principle of public bodies depending as little as possible on private resources.

79 Significantly, as happened with the general secretary, for a long time the personnel were Swiss nationals.

would hold the same relations with the Bureau Head as with the heads of telegraphs in all other member states. Furthermore, these two figures would be "in a completely different position from each other and will be placed, each dealing with what concerns him, under the direct orders of the Federal Councilor in charge of the Higher Direction of Telegraph Affairs"[80]. Because of the independence of these two figures, Curchod was invited to resign as Head of Swiss Telegraphs before taking up the position of Bureau Head. On 31 December 1868, Charles Lendi replaced Curchod at the head of Swiss Telegraphs and, from 1 January 1869, he became the first secretary general of what is today known as the International Telecommunication Union.

In another letter the Post Department sent to Curchod before the official start of the Bureau's activity, dated 1 January 1869, it was yet again repeated that the head of the new international office was dependent on the Swiss government, while his field of activity was extended. He would correspond directly with the various telegraph administrations, while "the Post Department on its side will monitor the necessary progress of the service and the use of the funds attributed to it"[81]. Control over the budget, control over the personnel, control over the Bureau's performance. By overseeing the office that had to regulate European telecommunications, the Swiss Post Department, and therefore the Swiss government, was in a position to wield great influence over them.

80 «dans un position complètement distincte l'une de l'autre et seront places tous deux, chacun en ce qui le concerne, sous les *ordres immédiats du Conseiller fédéral* chargé de la Direction supérieure des affaires télégraphiques». ITU-Corr, feuille n. 1/3, 14 October 1868, [*our italics*].

81 «Le Département des Postes de son cote surveillera la marche nécessaire de ce service et l'emploi des fonds qui lui sont attribués». *Lettre écrite par le Département des Postes adressée à Curchod*, 22 Décembre 1868, in Swiss Postal Archives, GDTT, T00A, 0059.

Chapter 5
Towards Rome Conference: Moves and Counter-moves (1868–1872)

Introduction

When the Bureau started its activity, the political setting in Europe had undergone considerable transformation. Up to the 1868 Vienna Conference the Telegraph Union had had a helping hand from the stimulus provided by Napoleon III's France, which was aiming at reviving its role of top continental power by means of very intense international political activity. From the late sixties, however, Napoleon III's fortunes began to decline and in 1867 the election victory of the republican opposition forced him to curb his expansionistic policy and concentrate more on domestic reform[1]. In the same year, he was forced to abandon Maximilian of Hapsburg to his destiny and consequently lose any influence over Mexican affairs[2].

The Second Empire collapsed a few years later with defeat in the Franco-Prussian War (1870–1871) leading to the German occupation of Paris, Napoleon III's exile and the establishment of the Third Republic[3]. The outcome of the war did not only give the newborn German Empire hegemony in Europe, but also notably re-enforced the role of Italy, which was able to round off its unification by annexing Rome – thanks to Napoleon's weaknesss and inability to go on defending the Papal States.

1 G. Antonetti, *Louis-Philippe* (Paris: Librairie Arthème Fayard, 1994), 278–280.
2 É. Anceau, *Napoléon III. Un Saint-Simon à cheval* (Paris: Tallandier, 2008), 410–412.
3 F. Roth, *La guerre de 1870* (Paris: Fayard, 1990).

Prussia's victory in the war drove the southern German states to open negotiations in order to join the Northern German Confederation, then completed in January 1871. The German Empire, born of Prussian initiative in open contrast with Austria, gave life to a powerful new political subject that would condition the destiny of Europe for the rest of the nineteenth and a large part of the twentieth century[4].

Great Britain was politically and economically the most important nation in Europe as well as the country where the telegraph had been most developed. In the late sixties and early seventies, the Prime Minister William Gladstone made a liberal turn and promoted a series of reforms tending to extend the vote and schooling, so putting an end the patronage system and furthering the democratization of society[5]. One of his moves was to nationalise the telegraph service, which up to then had been run exclusively by private companies.

The consequences of these great changes on the European political and economic chessboard were reflected in the TU and emerged very clearly during the Rome Conference of 1872.

First, the French delegate, who had previously carried out a leading role together with the Belgian and Swiss delegates, kept a low profile. While on the contrary Meydam, the delegate from the North German Federation and no longer Prussia alone, stonewalled, just as Jagerschmidt had done in Paris and Vienna. On numerous occasions, he tried to change some important articles in the convention, often clashing with Vincent and Curchod, who were more concerned about defending the *status quo.* Clearly, Meydam's actions were guided more by the need to command respect for the new political role of the Empire than real technical reasons.

In the second place, in nationalising its telegraph service Great Britain acquired the right to take part in the TU as a full member. Its arrival brought up the question of defining precisely who exactly could be considered a Union member. Was it the telegraph administration or

4 M. Kitchen, *A history of modern Germany: 1800 to the present* (Chichester, West Sussex and Malden, MA: Wiley-Blackwell, 2012), in particular chapters 6–8.

5 I. St. John, *Gladstone and the logic of Victorian politics* (London & New York: Anthem Press, 2010).

the state that controlled it? Up to then the question had never risen because each state possessed only one such administration. Great Britain came in, however, as an industrial and colonial power and with as many telegraph administrations as colonies. There was also the problem of historical precedence, since a British delegation had represented the British-Indian administration and thus was a member to all effects. It was no minor problem because in its specific political and historical position, Great Britain was the only nation to have two votes (with both a British and Indian delegation). The new status acquired by Great Britain, which enjoyed an objective technological advantage over other nations, put in serious difficulty the political equilibrium of the Union.

Thirdly, it was a period in which there was an explosion in building and laying long submarine cables. After the placing of the first transatlantic cable to function successfully for a sufficiently long period (1866), the mostly British private companies set up numerous intercontinental links, so allowing Europe direct contact with all the other continents in the system. The development of intercontinental communications made it essential to extend the Rome invitation also to delegates from the private companies running the international telegraphic service. However, the Union was born as an organization of states and though private companies lacked the right to vote, their presence alone caused friction and made new problems emerge. Friction was increased by the fact that the private sector managers were all British, so that a greater weight of private companies could cause the political centre of the TU to veer towards Great Britain. A further consequence of the changes in the international telegraph network and in general of communication flows on a planetary level was the need to extend membership to distant but already industrialized or industrializing countries like the States and Japan. The latter was already represented without voting rights in Rome, while the needs of the very private American telegraph service were discussed only marginally.

Finally, the new weight of Germany and Great Britain within the TU began to cast doubts on the Swiss strategy controlling the Bureau. Significantly, Germany presented at Rome a project intended to remove the new international office from Swiss control.

5.1 Replacing Curchod

In its first year of activity, the Bureau dealt with many questions concerning the ordinary management of the international telegraph network – from the languages to use in the telegrams, the nomenclature of the telegraph offices, the building of new lines, statistic tables, a survey on the role of women in telegraph offices to the editing of the *Journal Télégraphique.*

Relations between Curchod and Lendi were particularly proactive. The Bureau chief and the equal ranking Swiss Telegraphs head were under the authority of the Swiss Post Department and therefore of the Federal Council. The most pressing question in this first year was the routes to take and the tariffs between Europe, India and China. In January 1869, Lendi wrote to Curchod specifying that two different routes could be taken for telegrams from Switzerland to Egypt, either via Turkey (for 24 francs) or the new Malta line (32 francs). The latter was, however, much faster and, according to Lendi, Swiss companies preferred it[6]. Beyond the rule of the cheapest route established at Paris that telegraph operators had to follow on sending the telegram, users could indicate precisely which route they desired for their telegrams. It was, according to Curchod, a solution (he used the term "arm") to adopt with extreme care, but always leaving the user the possibility of bypassing some parts of the network for "political considerations or others exogenous to the service […] or for reasons of greater security or speed"[7]. With the delays that often occurred on the Turkish lines most Swiss users (generally businesses) preferred an almost 30% dearer tariff and sent their telegrams along the submarine cables

6 *Lettre écrite par Lendi adressée à Curchod*, 19 Janvier 1869, in PTT, GDTT, T00A, 0059.

7 «considérations politiques ou autres étrangères au service […] ou par motif qu'au plus de sûreté et de rapidité qu'elle offre aux correspondances». *Lettre écrite par Curchod adressée à Lendi*, 27 Février 1869, in PTT, GDTT, T00A, 0059.

passing via Malta[8]. Like other exchanges between the two heads the above illustrates the typical Swiss approach to telegraphy. One of the main interests of the Federal Council was in fact to allow national businesses to compete on an international level, so equal access conditions to the networks was essential, as the two following episodes show. In March 1869, Lendi brought to Curchod's attention an explicit request from a Swiss business corresponding regularly with the Indies[9]. It complained that unlike the British, the Swiss counted as two distinct words some compound monetary terms like "halfpenny". They judged it unfair competition because British companies saved considerable sums in sending telegrams very often full of such terms. Curchod answered that the Swiss government would do as the British did[10]. In July 1869, Lendi reported another complaint to the Bureau, again filed by a famous Swiss company of the time, Volkart of Winthertur, about an irregularity on the international line. The telegrams of Reuters News Agency transiting through Turkey on their way to India were given a priority over the others[11]. It was another case of unfair competition and a few months later Curchod wrote that he had taken up the matter with the Turkish Administration, which had answered that it was not alone in allowing priority to some telegrams, but only did so in exceptional cases. The practice was however, according to Curchod, contrary to the revised convention and he invited Lendi to report any further abuses[12]. The two above episodes show how international telegraphs were an essential instrument of work for Swiss companies and how they expected to receive the same treatment as other

8 *Lettre écrite par Curchod adressée à Lendi*, 21 Janvier 1869, in PTT, GDTT, T00A, 0059.

9 «maison de commerce suisse qui a des relations télégraphiques assez fréquentes avec les Indes». *Lettre écrite par Lendi adressée à Curchod*, 9 Mars 1869, in PTT, GDTT, T00A, 0059.

10 *Lettre écrite par Curchod adressée à Lendi*, 29 Avril 1869, in PTT, GDTT, T00A, 0059.

11 «priorité sur les autres». *Lettre écrite par Lendi adressée à Curchod,* 22 July 1869, in PTT, GDTT, T00A, 0059.

12 *Lettre écrite par Curchod adressée à Lendi,* 18 Septembre 1869, in ITU-Corr, feuille n. 62/34, 1869.

nationalities. On the question of equal rights, they found a listening ear and marked attention in the Federal Council, which was quick to raise matters with the Bureau. It is interesting to note that with the help of government mediation, companies were able to bring to the light broader problems and so themselves influence the evolution of the network.

The Bureau's first year of life was troubled by the resignation of the secretary Bellaroche for health reasons[13] but even more by that of Curchod, the mastermind behind the International Office. He stepped down both because of the prestige of the new job offered him (as head of the *Société du Câble Transatlantique Français*), and for purely economic reasons. In his 1869 managerial report, he detailed his financial circumstances, noting how the salary of only 8,000 francs, decided by the Federal Council "to respect the hierarchy" between the Head of Post Department and a subordinate in Curchod's position, was not enough. It was considerably lower than what had been recommended by the Vienna Conference, which had suggested 12,000 francs a month "as a minimum" for the head of the new international organization[14]. After having had the merit of envisaging and realizing his project, Curchod most probably expected better economic treatment, as is witnessed by the fact that in spite of the embarrassing nature of the matter he did not desist. Whatever, from 1 January 1870 Curchod went private and took over as head of the French company. It was very prestigious and highly paid position, given the big money circulating in transatlantic cable companies at that time[15].

---

13 The Federal Commission accepted his resignation in August 1869, reserving the right to see to a replacement, showing substantially it was taking the main decisions about the personnel in the International Office.

14 «pour maintenir l'ordre hiérarchique des traitements», «comme minimum». *Documents de la conférence télégraphique internationale de Vienne*, p. 456. See also *Rapport de gestion du Bureau international pour l'année 1869*, 31 Décembre 1869, in ITU-Corr, feuille n. 1/121, 1869. The problem of the head of the Bureau's salary was often a subject of controversy and Curchod's successor (Lendi) would have the same problems with the Federal Council.

15 See P. Griset, *Technologie, entreprise et souveraineté: les télécommunications transatlantiques de la France, 1869–1954* (Lille: A.N.R.T. Université de Lille III, 1993).

With the newyear, therefore, the Federal Council found itself with the delicate problem of Curchod's succession. It moved with extreme caution, which would turn out to be so decisive at the Rome Conference. The heart of the matter was that Curchod had not been nominated by the Federal Council, but by the Vienna Conference. Formally, therefore, the Council had no authority to appoint a new director and the decisionary power lay with the Conference. Yet the Post Department saw itself forced to adopt the necessary measures so that the service placed under its care would not suffer[16] and, even in an *impasse* like this, the Bureau would go on functioning. The issue was the lack of guidelines from Conference's decisions, though the Federal Council knew how move with the necessary tact. Therefore, while regretting no longer being guided by the enlightening evaluations of the delegates of the contracting States, the Federal Council would strive to carry out the mission entrusted to it in conformity with the Conference's views[17] and that was managing the Bureau. Formally, the Federal Council decided not to go ahead with fully replacing Curchod and on 9 December 1869 nominated Lendi director *ad interim*, actually placing all the Bureau's personnel under the higher authority of the Director of the Swiss Telegraph Administration[18]. As Lendi himself held, it would be up to the Florence Conference (then transferred to Rome as the new capital of the Kingdom of Italy) to decide both on the replacement of the director and, above all, whether to keep the Bureau going, a service which had proved itself useful while respecting the sovreignty and freedom of the individual countries over telegraphy.

16 «le service placé sous son autorité ne reste pas en souffrance». *Lettre écrite par Dubs adressée au directeur général des télégraphes,* 6 Novembre 1869, in PTT, GDTT, T00A, 0059.

17 «tout en regrettant de n'être plus guidé par les appréciations éclairèes des délégues des Etats contractants, le Conseil fédéral s'enforcera de remplir, conformément aux vues de la Conférence, la mission qui lui a été confiée». *Ibidem.*

18 «sous l'autorité supérieur du Directeur de l'Administration des Télégraphes suisses». *Projet pour le Budget du Bureau international pour l'année 1870 adressée au Département des Postes*, 10 Janvier 1871, in ITU-Corr, feuille 3/1, 1871; see also *Lettre écrite par le Département des Postes adressée à Lendi*, 31 Janvier 1869, in ITU-Corr, feuille n. 1/120, 1869.

Figure 14: Charles Lendi: Head of Swiss Telegraphs and *ad interim* General Secretary of the Bureau in 1869, then permanent from 24 May 1872 to 12 January 1873.

Source: ITU Archives.

In spite of the new appointment, Lendi did not leave his position as head of Swiss Telegraphs formally, but a second head, Frey, was drafted in so he could dedicate himself entirely to running the Bureau. The Federal Council, however, aimed to show its caution and that Lendi was not in charge to all effects and purposes.

In the memorandum of 13 December 1869, which was quoted during the Rome Conference, the Swiss Post Department asked all TU member states to express their opinion on how to proceed with the temporary replacement of Curchod. There has come to light in the archive an answer from the Head of Austrian Telegraphs Brunner, who followed the policy of reciprocal Austrian-Swiss support going back before the creation of the Union and so gave his "complete agreement to the dispositions that the Federal Council took over the re-organisation of the International Bureau"[19]. Thus, the Federal Council received at least one show of authorative support – important, too, in sight of the pending Rome Conference.

5.2 The Bureau as a peacemaker

The Special Commission proposed by France back in Paris in 1865 and created three years later at Vienna, met only once in all its history. It was convened in Bern in September 1871, with the aim of solving the awkward question of communications between Europe and the East[20].

As already mentioned, communications with India and China were a question of strategy for various European states, and in recent years newly opened telegraph lines had completely changed the situation. India for example, could be reached from Europe via three routes. The first traditional one transited through Turkey and was formed only of public landlines. The least efficient option, the tariff from London to Bombay cost 71 francs. The second route, which crossed Russia, had recently been sold by various public admnistrations to the *Indo-European Telegraph Company*, which had put up the charge to 112.50 francs. Then there was a brand new route with a junction in Malta, which was

19 «entier assentiment aux dispositions que le Conseil fédéral a bien voulu prendre à l'égard de la réorganisation du Bureau International». *Lettre écrite par Brunner adressée au Département des Postes*, 4 Février 1870, in ITU-Corr, feuille n. 1/5, 1871.

20 On the development and debates of the Special Commission called in Bern see *Procés-verbal des réunions de la Commission convoquée a Berne par les Administrations Austro-Hongroise pour le règlement des tarifs des Indes et de la Chine*, Imprimerie de C.-J. Wyss, Berne 1871 (SFA, E52, 514). Another interesting source is "Commission télégraphique spéciale de Berne," *Journal Télégraphique*, 1/23 (1871): 355–356 and "Commission télégraphique spéciale de Berne," *Journal Télégraphique*, 1/24 (1871): 373–374.

formed of long sections of submarine cables run by four different private companies with a London to Bombay tariff also at 112.50 francs.

To communicate between Europe and China, there were just two routes. The first started in Great Britain and went submarine to Russia and from there to China. It was a mixed line with sections run by public administrations and private companies, with a tariff of 107 francs. The second route could choose either of the two India options and then connect up with China by means of the *British-Australian Company*. The cost of a London-Hong Kong telegram was 175 francs.

Figure 15: "Telegraph Map of the Eastern World" published in the *Illustrated London News*, 8 July 1865.

Source: <http://distantwriting.co.uk/telegraphmap.aspx>.

The Special Commission did not turn out to be very efficient because it had little coercive force, and in fact as recounted below, it was suppressed during the Rome Conference. The interests at stake were enormous. Some states did not want to lose their importance for through-traffic and the consequent income in the international telegraph network. Other countries had economic interests and rich commercial exchanges with the East and aimed for an efficient service as cheap as possible. Then there were the private companies, especially submarine cable ones, which had invested enormous sums of money in the construction of the underseas network and wanted to fix their own tariffs. Surrounded by these interests, the Bureau offered itself as the main agent of conciliation, the function that, in a certain sense, it had been created for. Already in this meeting, which significatively took place in Bern as Austria had requested, the Bureau played an active role in the debate and "As a conclusion to this communication, it could take the liberty of advising, as a concilitary solution and subject to legal considerations accepting in actual fact until the next conferences or the next meeting of a special commission"[21]. This would be the solution the Commission adopted in the end.

5.3 Briefing the Swiss delegate(s) for Rome[22]

As in every conference, the Federal Council gave its delegate instructions to follow. Lendi, who was to take part with the dual function of Head of Swiss Telegraphs and Director *ad interim* of the Bureau, had to defend as far as possible the directions taken in the Paris and Vienna conferences. In the case of new provisions being introduced, he was asked to act so that they were "the most liberal and favourable to international traffic".[23] Secondly, articles 32, 33 and 34 were to be maintained operative in their essential points and yet again he was to act so that the individual countries continued to enjoy a certain freedom in

21 «Comme conclusion de cette communication, il crut pouvoir prendre la liberté de conseiller, à titre de solution conciliatrice et sous toute réserve de la question de droit, l'acceptation de fait jusqu'aux prochaines Conférences ou jusqu'à la réunion d'une Commission spéciale». *Procés-verbal des réunions de la Commission convoquée a Berne*, p. 17.

22 All the quotations in this section are taken from *Projet Conférences de Rome. Instructions*, 1 Décembre 1871, in SFA, E52, 515.

23 «les plus libérales et les plus favorables au trafic international».

reducing tariffs. Thirdly, Lendi was to oppose word reduction in a simple telegram from 20 to 10. Other instructions concerned the income from transit telegrams on Swiss territory. He was, in fact, authorized to contract tariff reductions providing "the reductions thus obtained will be such as *to attract* to a certain point *transit over Swiss territory*"[24]. He was also requested to manoeuvre so that the new line connecting Germany to Italy crossed over Swiss territory. Lastly, he was told: "The delegate will act to maintain in Bern the International Bureau of Telegraph Administration, but will refrain from making propositions about its organisation"[25] and sent other observations on the matter.

It was not Lendi, however, who acted as Swiss delegate at Rome, as an operation forced him to stay in bed the whole length of the Conference. The Bureau was represented by the secretary, while Lendi put forward as Swiss representative the name of Curchod[26], who was already present at the Conference as the delegate of his private employers. Lendi's proposal was warmly welcomed by the Federal Council, and. Curchod's presence and personality were probably decisive in warding off Germany's attack on Switzerland's right to manage the Bureau.

5.4 The German move to snaffle the Bureau

In early December 1871, practically at the same time as the opening of the Rome Conference, Germany presented a long memorandum then transformed into a proposal for an amendment. The core of the German proposal was an alternative to the current model of the Bureau, in order

24 «les réductions ainsi obtenues aseront de nature à attirer dans une certaine mésure *le transit par le territoire Suisse*». [*Our italics*].

25 «Le délégué agira pour le maintien à Berne du Bureau International des Administrations télégraphiques, mais il s'abstiendra de faire des propositions relatives à son organisation».

26 *Lettre écrite par Lendi adressée à la Direction des télégraphes*, 4 Décembre 1871, in SFA, E52, 515.

to remove it from a single country and/or its postal departments, and make it an independent body that "would answer to the Conference alone"[27]. Though Switzerland was mentioned only in a positive way, the proposal was a direct attack on the power it had acquired over the Bureau.

The Germans listed three possible methods for running the International Bureau. One provided for a conference-appointed head placed under the administration of the Telegraphs Department of the state hosting the most recent conference. This meant he would be a kind of secretary for the Post Minister of the host country and would be unable to subtract himself from the influence and vested interests of the administration in question. To manage the Bureau, however, an independent figure was needed who "would follow only the interests of maintaining European unity"[28] and it would be difficult to find him if he had to change seats in continuation, without taking into consideration the cost of relocations weighing on the Bureau's budget. So this first proposal was rejected by the Germans, for reasons very similar to those put forward three years earlier in Vienna.

A second way of running the Bureau, which corresponded more or less to the present situation, provided for the appointment of an administration rather than a director. In Vienna the choice had fallen on Switzerland, but the Germans felt was not so much for its intrinsic qualities (the document quotes its central position and neutrality) as for the great esteem the then Head of Swiss Telegraphs, Curchod, enjoyed[29]. The Germans held the model contained a basic flaw which could no longer be tolerated, and which Switzerland itself had understood:

> Nevertheless all the doubts expressed above which call for *the independence of the official from all special administration* not being raised, and the justice of these considerations noted by the same, that the Swiss Federal Council, with its

27 «la création d'un organe permanent qui [...] ne sera responsable qu'à la Conférence seule». *Memorandum du directeur général des télégraphes de l'Empire allemand*, 10 Décembre 1871, in SFA, E52, 515.

28 «ne suivit d'autres intérêts que ceux du maintien de l'unité européen». *Ibidem*.

29 *Ibidem*.

most tactful awareness of its mission did not have the said employee replaced by another person, who after a short period of work, resigned the position.[30]

So the way Switzerland had acted after Curchod's resignation in avoiding taking responsibility for a permanent replacement and in appointing Lendi only *ad interim* highlighted the problem of the independence of the Bureau head from a national telegraph administration.

The Germans favoured their third option, initially also supported by the Belgians, which they felt could "alone satisfy all needs"[31]. It was to appoint the Bureau head in a plenary sitting and "subordinate exclusively to the Conference, its *consulting and executive authority*, which constitutes its tradition and *whose existence is closely linked to that of the Conference itself* so as to guarantee as far as possible the most precise execution of its duties"[32]. Once the head had been subtracted from the influence of a single telegraph administration, a state was needed to to act as supervisor, especially over accounts and the correct handling of funds. According to the Germans, though belied in part by the words of the Federal Council: "The Swiss Administration [...] will also assume responsibility in the future of the Higher Direction"[33].

In spite of being confined to bed, on 12 December 1871 when the conference was in full swing, Lendi wrote to the Swiss Post Department to ask for further instructions on this difficult matter, which he would

30 «Toutefois tous les doutes exprimés ci-dessus qui demandent *l'indépendance de ce fonctionnaire de toute administration spéciale*, n'en étaient point lèves, et la justesse de ces considérations se fait remarquer par cela même, que le Conseil fédéral de la Suisse, avec une connaissance de sa mission *pleine de tact*, n'a pas fait remplacer par un autre personne le dit employé qui, âpres une courte durée de ses fonctions, avait quitté sa position». *Ibidem*, [*our italics*].

31 «seule satisfait à toutes les exigences». *Ibidem.*

32 «subordonne exclusivement à la Conférence, en est *l'autorité consultative et exécutive*, en constitue la tradition, et dont *l'existence est étroitement liée à celle de la Conférence elle-même*, de manière à garantir, autant que possible, la plus exacte exécution de ses obligations». *Ibidem*, [*our italics*].

33 «L'administration suisse [...] se chargera sans doute aussi à l'avenir de la direction supérieure». *Ibidem.*

then discuss with Curchod. He felt the German proposal appeared to want to give more power to the Director of the Bureau.

> [...] but the arguments put forward for the suggested modifications are more or less of the type to prove that the *independence* of the Directeur and his freedom of action are as complete with *the present organisation* as they would be after the inclusion of the project, and I do not doubt that in the moment in which the Articles [...] are debated, I could replace myself with myself and easily convince the members of the Conference and I think that it will be more ready to keep the present positions while *reserving the right to express to the Federal Council its wishes over the choice of the Director.*[34]

Lendi seemed quite sure that Switzerland could easily repel the German attack by simply illustrating the worthiness of its performance up to then. He trusted in the fact that the other states would accept that in the future too the Federal Council would cast its own vote and therefore preference for a candidate. However, in the case of the German proposal being approved, Lendi requested instructions over what to do and particularly "if the Federal Authority would think of being able to keep the Higher Direction of the International Bureau"[35]. In other words, if the Bureau were to be removed from Swiss influence and no longer under its control, was Switzerland still interested in keeping the Bureau on its territory?

A few days after receiving this letter, the Federal Councillor Head of the Post Department (Challet-Venel) sent a detailed report to the Council beginning with Curchod's resignation and the fact that:

34 «mais les arguments mis en avant pour les modifications proposées sont plutôt de nature à prouver que l'*indépendance* du Directeur et sa *liberté d'action* sont aussi complètes *avec l'organisation actuelle* qu'elles ne le seraient d'âpres le projet ci annexe et je ne doute pas qu'au moment ou les Articles [...] seront discutés, je puisse, moi au moi remplaçant, facilement convaincre les membres de la Conférence et je pense que celle-ci sera plutôt disposée à maintenir les dispositions actuelles en se *réservant de faire exprimer au Conseil fédéral son vœu sur le choix du Directeur». Lettre écrite par Lendi adressée au Departement fédéral des Postes*, 12 Décembre 1871, *ibidem*, [*our italics*].

35 «si l'Autorité fédérale penserait pouvoir conserver la Direction supérieure du bureau international». *Ibidem.*

> the Federal Council will not deem it necessary to proceed to nominating a permanent replacement for the outgoing director [...] in an obvious attempt to understand and reserve for the future Rome Conference all latitude for confirming the temporary state.[36]

This move could be read in two ways. On one hand, it is another example of the caution and tact with which the Federal Council moved in a delicate situation that could have compromised its control over the Bureau. On the other hand, more artfully, the Swiss decision can be perceived as an attempt to repeat the strategy that had already been effective in the case of Curchod, i.e. in suggesting that the Head of Swiss Telegraphs was the natural candidate for the chair of the Director of the Bureau. It was clearly no imposition but an attempt to trace a guideline, hoping naturally in the approval of the future conference. Approval arrived, the temporary mechanism Switzerland had suggested became the usual procedure and, until WWII, the secretaries general of the TU (and later the International Union of Telecommunications) would all come from Swiss telegraphs.

The German plan, however, substantially changed the cards on the table, introducing some elements which Switzerland could not have accepted. Firstly, the fact that "it will no longer be the Administration of the country chosen as seat of the International Bureau which will elect the Bureau Head and personnel"[37]. Switzerland evidently wanted the power to nominate the Director and his staff. Secondly, the Post Department complained that this plan was too costly, for Switzerland had always found it important to keep the Bureau's expenses down. The head's salary would be increased from 12,000 to 20,000 francs, while he would have access to another 30,000 francs for staff. Challet-Venel felt the main aim of this plan was to "increase the competence of the International Bureau's Head and deliver his decisions from the control

36 «le Conseil fédéral ne jugea pas devoir procéder à une nomination définitive pour remplacer le Directeur démissionnaire [...] dans le but facile à comprendre de réserver a la future conférence de Rome toute latitude fait pour la confirmation de l'état provisoire». *Lettre n. 140/30 écrite par Challet-Venel adressée au Conseil fédéral à Berne,* 18 Décembre 1871, *ibidem.*

37 «ce ne serait plus l'Administration du pays choisi comme siège du Bureau International qui aurait à élire le Directeur et les employées du Bureau». *Ibidem.*

of the Higher Administration of the country seat of the Bureau. It is precisely the opposite of what was desired in 1868"[38]. He saw the main aim underlying the German proposed amendment was to remove the control of the Bureau from Switzerland and the fact that Switzerland could be given a role of control over the accounts was only an empty formality[39].

The implementation of this plan would have had disastrous consequences. In the first place, the director nominated in this way would be the "executive power of the conference: he will no longer be [...] a simple agent"[40]. This authority would be going well beyond the ideas and intentions of the various governments who had given life to the Bureau in Vienna, and according to the Federal Council would question the national sovereignty of individual countries.

> This would allow the International Bureau to enter *a new sphere of action* and give the bureau *an authority that seems incompatible with the dignity and freedom of contracting telegraph administrations* [...] the general agent using sovereign authority in the interval would not take long to become an obstacle for the international service.[41]

It was felt that one consequence of approval of this plan would be a rejection of everything decided at Vienna, where the conference had

---

38 «augmenter la compétence du Directeur du Bureau international et *d'affranchir ses décisions du contrôle de l'administration supérieure du pays au se trouve le Bureau.* C'est précisément le contraire de ce qu'on a voulu établir en 1868». *Ibidem,* [*our italics*].

39 «vaine formalité». *Ibidem.*

40 «pouvoir exécutive de la conférence: il ne serait plus [...] un simple agent». *Ibidem.*

41 «Ce serait donc faire entrer le Bureau International dans une *nouvelle sphère d'action* et donner à ce bureau une *autorité* qui parait *incompatible avec la dignité et la liberté des Administrations télégraphiques contractantes* [...] l'agent général usant d'autorité souveraine dans l'intervalle ne tarderait pas à devenir un obstacle pour le service international». *Ibidem,* [*our italics*]. Interesting to note how the Swiss government used on this occasion the same words adopted four years earlier by the French delegate at Vienna. While in Vienna "the dignity and freedom" of the states could be endangered by Curchod's project for the Bureau, it was the German project which aroused the disdain of the Swiss in Rome. Maybe, in order to find agreement with the French, the Swiss delegates re-directed the accusation that had been made against them on the Germans, questioning the liberal content of their proposal.

thought of a completely different model of the Bureau – a body with an essentially consultative power, functioning above all at an administrative level in order to solve the everyday problems concerning the administrations. For this reason:

> This body must be attached to a higher administration chosen by the conference and function under the direction of this administration in order to be at the centre of the everyday matters and be able to use the special resources that the various branches of this administration can justify.[42]

The Swiss delegate therefore was instructed to oppose the German plan, though keeping mind that it could be approved, given that he represented a small country and "cannot pretend that in a conference gathering the delegates of a large number of States that Switzerland could impose its viewpoint"[43]. For this reason, diplomatically "it is best to authorize our delegate to declare that the Federal Council *in all cases* is willing to help with the new organisation"[44].

On the same day, Challet-Venel wrote a letter to Lendi which discloses other details about the German plan. First of all, the head of German Telegraphs had not taken the Swiss Post Department unawares because "it is the repetition of an idea that I have heard being developed very ably by one of the influential members of the Telegraph Commission which met in Bern last autumn"[45]. It was a matter, Challet-Venel

42 «Cet organe *doit être rattaché à une administration supérieure choisie par la conférence et fonctionner sous la direction de cette administration* afin de se trouver *au milieu des affaires courantes* de chaque jour et de pouvoir utiliser les ressources spéciales qui peuvent prouver les diverses branches de service de cette administration». *Ibidem*, [*our italics*].

43 «nous ne pouvons pas prétendre que, dans un conférence qui réunit les délègues d'un grand nombre d'Etats, la Suisse puisse imposer sa manière de voir». *Ibidem*.

44 «il conviendrait d'autoriser notre délègue a déclarer que le Conseil fédéral, *dans tous les cas*, prêtera volontiers son concours à la nouvelle organisation, dans la mesure qui serait réclamé dans l'intérêt de toutes les administrations télégraphiques contractantes». *Ibidem*.

45 «c'est la reproduction de l'idée que j'ai eu l'occasion d'entendre développer très habilement par un des membres influents da la Commission télégraphique, réunie a Berne l'automne dernier». *Lettre écrite par Challet-Venel adressée à Lendi*, 18 Décembre1871, *ibidem*. In the Special Commission minutes there is, however, no

continued, of two currents of thought which anyway already been expressed in Vienna. On one side, the concept of the Bureau as a sort of "executive power of the Conference, a sort of dominator"[46] with a great decisional freedom running the risk of creating a supranational power. On the other hand, a body working in agreement with all countries, leaving freedom of expression and that, according to the Swiss Post Department, has already been experimented and has shown it is "an instrument that has well functioned so that there (is) no need to try an experiment in the opposite direction"[47].

In a long and interesting final passage, which needs quoting in full to give a complete illustration of the Swiss strategies, the Post Department attacked the German plan harshly and went as far as to hold that, if Switzerland could not nominate the director or wield full power over the Bureau, it might well give up hosting the body on its territory. This document reveals most clearly that the interest of the Swiss government was not so much in having the Bureau on its territory as being able to manage its operations, decisions and the choice of directors, etc.

> I must say that I do not really understand the reason for not going still further than what is proposed by the General Director of Imperial Germany's Telegraphs. Since the Higher Administration of the country chosen as the seat of the International Bureau would not be in charge of the Bureau's organisation […] it seems to me derisory to leave to this higher administration […] *the somewhat irksome task of establishing the accounts and deploying the funds* […] In this manner […] the dignity of the higher administration would be managed better, whichever one, whether of Switzerland or another State, rather than asking the Administration to make a certainly insignificant intervention, to ask for no intervention at to be exercised over the International Bureau. If Mr de Chauvin's proposition were to find favour with the Majority of the Members of the Conference, which I am tempted to doubt, it would be better to complete it in the sense indicated above and suppress any intervention from the Administration of the country hosting the International Bureau. If on the contrary, without going so far the Conference would prefer to keep the

---

trace of this idea, which must have circulated in an informal way among the higher decision-making spheres.

46 «pouvoir exécutif de la Conférence, un sorte de dominateur». *Ibidem.*

47 «l'expérience a prouve des lors […] que l'instrument a aussi bien fonctionné pour qu'il ne doit pas (sic) nécessaire de tenter une expérimentation dans le sens contraire». *Ibidem.*

intermediary proposition as presented by Mr de Chauvin, *I could not really understand how the Swiss Administration could refuse to go on with its co-operation even in the diminished position assigned.* It seems to me that we will have to go to the end, [...] with the contracting telegraph Administrations, and that we will have to submit to [...] the common interest [...] you should in the name of the State you represent bring up all the inconveniences and dangers of the new system which they are trying to make the organization of the International Bureau and oppose strongly and with great conviction that article 61 be modified in the sense proposed.[48]

Additional instructions for Lendi were attached to the above letter. The Federal Council wanted to emphasize that it would accept the Conference's decisions over changes to article 61, but it also emphasized that the proposed changes would not improve the organization of international telegraphy, therefore inviting its delegate to oppose them resolutely[49].

---

48 [...] je dois dire que je ne comprends pas bien par quel motif on n'irait pas encore plus loin que ne le propose le Directeur Général des Télégraphes de l'Empire Allemand. Du moment que l'administration supérieure du pays choisi comme siège du Bureau International ne serait plus charge de l'organisation du Bureau [...], il me parait *dérisoire* de laisser a cette administration supérieure [...] *la charge aussi fastidieuse de contrôler l'établissement des comptes et des emplois des fonds.* [...] De cette manière, on ménagerait mieux [...] *la dignité de l'administration supérieur*, quelle qu'elle puis se être, que ce soit celle de la Suisse ou de tel autre Etat, en renonçant tout à fait à faire appel à aucune intervention, plutôt qu'en ne faisant à cette Administration qu'une pour insignifiante dans la surveillance à exercer sur le Bureau International. Si la proposition de Mr de Chauvin devait trouver quelque faveur auprès de la Majorité des Membres de la Conférence, ce dont je suis tenté de douter, il vaudrait mieux la compléter dans le sens ci-dessous indiquée et supprimer tout intervention de l'Administration du pays ou siégerait le Bureau International. Si, par contre, la Conférence, sans vouloir aller aussi loin, préférait d'en tenir à la proposition intermédiaire telle qu'elle est présentée par Mr de Chauvin, *je ne venais pas trop comment l'Administration Suisse pourrait refuser la continuation de son concours, même dans la position amoindrie qui lui serait ainsi assignée.* Il me semble que nous devrons d'aller jusqu'au bout, en fait de bon pouls envers les Administrations télégraphiques contractantes, et que nous devrons nous plier à ce quelles [...] l'intérêt commun. [...] vous devrais au nom de l'Etat que vous représentes faire ressortir tous les inconvénients et les dangers du nouveau système dans lequel l'on prétend faire entrer l'organisation du Bureau International et vous opposer fermement et avec pleine conviction, à ce que l'article 61 soit modifiée dans le sens propose». *Ibidem.*

49 *Instructions supplémentaires*, 18 Décembre 1871, *ibidem.*

5.5 The Germans at work

The Third International TU Conference opened on 1 December 1871 with 21 states taking part, including both Great Britain which had nationalised its telegraphs between 1868 and 1871 and representatives from private companies. The presence of the latter made the Rome Conference far different from Paris and Vienna. After a heated debate in the first session, it was decided to admit the representatives from the private companies to all sessions but with no right of vote. Whatever, the presence alone of the private company representatives guaranteed they had the chance to propose new norms and modifications[50]. The private companies were pro-active, especially over tariffs and modalities for joining the Union. Their opinions set off a lively debate, taken up repeatedly during all the sittings, about whether the Convention should look after the interests of the private companies as well as those of the public administrations and their relative clients[51]. Curchod, present at the Conference right from the beginning, also took part in this debate[52].

A second element contributing to change the balance of power within the Conference and the exchange of views during the debates was the new political climate of the early seventies. For a start, Great Britain alone enjoyed a double vote, though indeed its delegates did not really exploit it to the full. Secondly, after its recent military and political defeats, France found itself unable to make any meaningful proposals or influence important debates, as it had in the past. Differently, the German delegate Meydam intervened systematically proposing amendments on many articles and attacking those of others, though his attention seeking was well contained by the Belgian Vinchent. While Meydam tried all the time to subvert the equilibrium created at Paris and Vienna by proposing substantial changes, Vinchent worked at the

50 *Documents de la conférence télégraphique internationale de Rome*, pp. 219–235.

51 See also Codding, *The International Telecommunication Union*, 26.

52 See in particular the discussion on articles 32 and 34 about tariffs and on article 66 about joining the Union. *Documents de la conférence télégraphique internationale de Rome*, especially pp. 262–274, pp. 382–402 and pp. 493–505.

defence of the *status quo*. Thanks also to his greater international experience and the esteem he enjoyed, he always managed to impose his point of view over Meydam and consequently carried out the role of mediator that had previously been Curchod's.

Dramatically Switzeland, which had had such an important role in Paris and Vienna, suddenly found itself without a delegate. As Head of Swiss Telegraphs, Lendi had been appointed delegate, but his lasting illness left Switzerland with no official representative for the first six sessions. Curchod was then brought in *in extremis,* by the Federal Council to represent its interests[53]. From then on the dynamics of the debate changed radically. Though the questions in the seventh session were on the whole of minor importance, Curchod took up the duet he had performed with Vinchent at Paris and Vienna. The two veterans at once sidestepped Meydam who lost momentum as the days went by. The German proposals became less impactful, while the delegates went back to treating Vinchent and Curchod as the two pivots important questions rotated around. Tellingly, all the most important proposals were discussed after Curchod became Swiss representative. They included lower prices for extra-European simple telegrams with fewer words (art. 32), the freedom to up or down tariffs as the private companies wanted (art. 34), the possibility of abolishing the antidumping norm on international lines (art. 34 last paragraph) and above all the definitive regulation of the Bureau (art. 61).

5.6 The Bureau left in Swiss hands

For Switzerland, the most important question discussed in Rome was keeping the Bureau in the hands of its Administration. The German delegation had proposed an amendment to article 61 which would have effected a major change in the Bureau's structure for at least three

53 *Ibidem*, p. 329.

reasons: 1) the head would have been designated by the Conference and would be responsible to it alone; 2) the head alone would have been responsible for the organization of the Bureau; 3) the administration designate (presumably Swiss) would be required to control only the accounts. Evidently, the German move was to subtract the Bureau from Swiss influence.

Before the amendments were presented, Curchod wrongfooted everybody by asking to read a memorandum from the Swiss Post Department. With this move, Curchod wanted to highlight both the Swiss Administration's desire to clarify article 61 and the correctness and skill with which the task had been carried out. The memorandum contained a summary of how the problem of replacing Curchod had been dealt with, i.e. Switzerland had "for the time being placed"[54] the Bureau under the direction of the Post Department, appointed Lendi Director *ad interim* without having him resign as head of Swiss Telegraphs. It was also added that the Bureau would go on functioning in a complete independence from the Swiss Telegraph Administration. This temporary solution had been adopted in order not "to compromise at all the decisions of the next Conference, which will be able to make on the subject in question without any thouht of the past any disposistions it deems appropriate"[55]. In concluding, the Post Department confirmed it was ready to submit to the decisions taken by the Conference, just as it had done in the past, and that, when it could have exploited the situation, it had opted for a temporary solution; nevertheless Curchod added in the conclusion a detail that almost sounds like a threat:

> In the case of residence being kept in Switzerland and conditions of organization changed, the Federal Council would reserve the right to *examine these new conditions before giving its consent.*[56]

54 «placé provisoirement». *Ibidem*, p. 482.

55 «ne préjuger en rien les décisions de la prochaine Conférence, qui pourra prendre sur le sujet en question, sans aucune arrière pensée, toutes les dispositions qu'elle jugera convenables». *Ibidem.*

56 «Le Conseil fédéral se réserverait, au cas où, tout en conservant à la Suisse la résidence de ce Bureau, on en modifierait les conditions d'organisation, *d'examiner ces nouvelles conditions, avant de leur donner son acquiescement*». *Ibidem,* [*our italics*].

Expressing in a diplomatic way the uneasiness the Federal Council felt, Curchod informed the assembly that if the Geman proposal passed, the Swiss government intended to decide if and how to keep the Bureau on its territory. This was a further step respect to the exchange of letters between Lendi and the Post Department. It is not clear how far Curchod himself was responsible for the harder line adopted, or how far he was following orders.

Curchod's intention during the debate had been clear. He wanted to show the correctness of the Swiss government and win the favour of the other administrations in order to reduce the importance of the proposed German amendment. To judge from the debate, his strategy was a perfect successful. Meydam made a timid intervention that proposed to keep the Bureau in Switzerland, and of all the changes suggested, the only one he presented was about the Conference appointing the director directly.

Figure 16: The technical delegates at the Conference of Rome.

Source: ITU Archives.

It followed that all the interventions made by the most imporant delegates were all in favour of Switzerland. The first to take the word was Vinchent:

> To satisfy the desire manifested to give the International Bureau a governmental character, M. Staring proposed in Vienna *to designate a State* instead of choosing a person. This solution, which was considered almost an expedient, *experience*

showed to be excellent. So M. Vinchent will not discuss the serious considerations brought forward by the German Administration [...] He will pay homage to the Federal Administration for the manner in which it has carried out its mandate and the results it has obtained, while safeguarding with the greatest discretion the financial interests of the states. He has no objection to make to the plan put forward by the German Administration; the only thing that militates in its eyes against all innovation is the success obtained by the present body. He will vote therefore for maintaining the present state of things.[57]

Once again, a key theme in the debate among technicians was practical experience, i.e. a solution could be considered valid if it had functioned in everyday practice[58]. He concluded by saying that given the tact and equilibrium with which Switzerland had managed the first years of the Bureau, there was no reason for modifying the resolution taken in Vienna.

Brunner followed him and added that "the creation of an International Bureau (could be considered) one of the best ideas to come out of the Vienna Conference and one of the most efficient guarantees of the union of the states"[59]. He also confirmed that choosing Switzerland

---

57 «Pour satisfaire au voeu quis'était produit de donner au Bureau international un caractère gouvernemental, M. Staring a proposé, à Vienne, de *désigner un Etat*, au lieu de choisir une personne. Cette solution que l'on considérait presque comme un expédient, *l'expérience a montré qu'elle était excellente.* M. Vinchent ne discutera donc pas les considérations sérieuses invoquées par l'Administration allemande. [...] Il rend hommage à l'Administration fédérale pour la manière dont elle a accompli son mandat et pour les résultats qu'elle a su obtenir, tout en sauvegardant, avec la plus grande discrétion, les intérêts financiers des Etats. Il n'a pas une seule objection à opposer à la combinaison proposée par l'Administration allemande; *la seule chose qui milite, à ses yeux, contre toute innovation, c'est le succès obtenu avec l'organisation existante.* Il votera, en conséquence, pour le maintien de l'état actuel choses». *Ibidem*, pp. 483–484.

58 In the previous conferences too technicians like Curchod repeatedly explained how in running a telegraph service, it was fundamental to learn from experience and that once a solution had been found there was no need for change. A typical approach of technicians could acquire political resonances in circumstances like these.

59 «M. Brunner regarde la création du Bureau international comme une des idées plus heureuses de la Conférence de Vienne et une des garanties les plus efficaces de l'union des Etats». *Ibidem*, p. 484.

as seat of the Bureau and entrusted the running to it had actually depended on the choice of Curchod as the Director. This, he felt, had been understood perfectly by the Swiss government too, which had taken no steps to replace him. The German proposal was in fact superfluous, given that the Swiss Administration itself had had the tact to wait for the conference to express its opinion before appointing an official successor to Curchod. As for the future, "one could not change the mode of nomination and limit oneself to getting the Federal Council to send *in a more or less direct form* the wishes of the Conference"[60].

What Brunner did request was the director's independence from the Swiss Post Department. However, it would have been of no use to take up the German proposal "specifying what the sphere of duties of the Central Bureau is, in noting its independence from the Telegraph Administration it works with and finally leaving the Head of the Service to nominate the staff he will employ"[61].

Following his old custom of directing the debate at will, Curchod drew on Brunner's remarks and suggested some changes in form he felt could dispel any doubts about Swiss Telegraphs interfering in the running of the Bureau:

> It is enough, to obtain this, to specify that the authority this service is placed under is the Higher Administration the telegraph service answers to in any way. [...] In this way will be satisfied the desire not to attribute to a telegraph Office a prevailing influence over the International Bureau, while leaving it under the strict surveillance and regular control of a state.[62]

60 «l'on pourrait ne pas changer le mode de nomination et se borner à faire parvenir au Conseil fédéral, *sous une forme plus ou moins directe*, l'expression des désirs de la Conférence». *Ibidem*, p. 484, [*our italics*].

61 «en spécifiant quelles sont les attributions du Bureau central, en constatant son indépendance à l'égard de l'Administration télégraphique auprès de laquelle il fonctionne et, enfin, en confiant au Chef de ce service les nominations du personnel qu'il emploie». *Ibidem*, pp. 484–485.

62 «il suffit, à cet effet, de spécifier que l'autorité sous laquelle ce service est placé est l'Administration supérieure dont relève également le service télégraphique. [...] De la sorte, l'on satisferait an désir de ne pas attribuer à un Office télégraphique une influence prépondérante sur le Bureau international, tout en le laissant sous la haute surveillance et l'autorité régulière d'un Etat». *Ibidem*, p. 485.

Given the hierarchy of the Swiss government, the change was however more formal than substantial. Control over the Bureau passed from the Swiss Post Department directly to the Federal Council, but since the Department Head was a member of the Federal Council and all the departments were on the same level, the Bureau would continue to be under the direction of the Swiss government.

Seeing himself in the minority, the German delegate asked the assembly to take a decision over the main point of his proposed amendment, if the Conference or the designated administration, i.e. Switzerland, was to nominate the Director. The result of the voting was a triumph for Switzerland and the cautious policy of its government. The German amendment was rejected by 15 votes to 3, while Curchod's version was "passed without comment"[63].

Switzerland did not only outmanoeuvre the attempt to remove the Bureau from its control, but it managed to acquire further power over it, ratified by the other delegates. While the Bureau had previously been formally subject to a technical control by the Post Department, now it openly passed under the control of the Federal Council. It was another international recognition for Switzerland, as well as being a leap in quality. Now the Federal Council could quite legitimately adopt a political strategy with the Bureau.

What is more, after the Rome Conference the powers attributed to the Bureau would be still greater, or rather the Bureau would be in a new position of strength. Indeed, the Special Commission's substantial incapacity to reach any conclusion was officially recognized, and after a lengthy debate Brunner suggested suppressimg it and adding a norm whereby in the case of serious controversy, six member states could ask for the next conference to be put forward. As Vinchent acutely observed, processes were being simplified. The elements forming the TU would now be reduced from three to only two (Vinchent spoke specifically of "gears") and "everything outside the boundaries of the Bureau's action will be within that of the Conference"[64]. No conference was ever

63 «admise sans observations». *Ibidem*, p. 487.

64 «tout ce qui sortira des limites de l'action du Bureau international rentrera dans le domaine de la Conférences». *Ibidem*, p. 481.

put forward, so that the suppression of the Special Commission made the Bureau the only body to settle questions arising between conferences and gave it a new power to intervene actively in the management of telecommunication networks.

5.7 Swiss reactions

The Rome Conference closed on 14 January 1872 and, on 19 March, Lendi sent the Post Department a report on the outcome. All in all, he praised Curchod's contribution and "the successes he obtained in defending Switzerland's interests"[65]. He specified that the defence of article 61, the main issue of the Conference for Switzerland, had been successful and that the Swiss government had been again designated[66] for the organization of the Bureau. He also cited a special report on the subject drawn up by Curchod which he would make available to the Federal Council when it desired to go ahead with the definitive organization of the Bureau[67].

As by now should be clear, despite the formal change in the body the Bureau would have to answer to, there was no real change in how decisions were taken. On 24 May 1872, for example, the Federal Council following suggestion of the Post Department named Lendi Director of the International Bureau from 1 June 1872[68]. Thus, exactly as before, the Federal Council acted after listening to the Post Department, which had suggested the nomination of its own man as director of the Bureau and had come out on top.

---

65 «les succès qu'il a obtenu en défendant les intérêts de la Suisse». *Conference de Rome. Rapport écrite par Lendi adressée au Département fédéral des Postes*, 19 Mars 1872, in SFA, E52, 515.

66 «de nouveau désignée». *Ibidem*.

67 This report has not however emerged in the archives consulted. It could well be an important document for understanding if and how Curchod managed to direct the debate.

68 «sous la proposition du Département des Postes». *Lettre écrite par le Département des Postes adressée à Lendi*, 10 Juin 1872, in ITU-Corr, feuille n. 1/4, 1872.

There was not much change either in the director's salary either, which provoked Lendi into taking a tough stance. In May 1872, he wrote to the Post Department and cited the Federal Council's decision that the salary of the new director would be on a par with Curchod's (8,000 francs). He felt it was unacceptable because the Rome Conference had established 8,000 francs a year for the secretary, so that a dependent would receive the same salary as the director[69]. All the Federal Council's answer was to lower the director's salary to 7,000 francs (29 May 1872). Yet again Lendi gave a harsh reply, attaching to his letter a copy of the Conference debate in which the delegates asked for the Director's salary to be raised: "The Conference [...] decided that the thanks it addressed to the Swiss Federal Council for the management of the International Bureau rather than express its desire to increase the financial position of the Director and Secretary of the Bureau will be put in the minutes"[70]. He felt the Federal Council was going against the Conference's wishes. It was not the first time, however that the Federal Council had decided not to employ the whole budget allocated by the Conference, though it had just been raised to a 50,000-franc ceiling. This time, however, the government went against the Conference's wishes, perhaps irritated by Lendi's reactions. It is just one of the examples of how after Rome the Swiss government felt itself firmly in charge of the Bureau and therefore free to run it as it desired.

69 *Lettre écrite par Lendi adressée au Département des Postes,* 28 Mai 1872, in ITU-Corr, feuille n. 1/2, 1872.

70 «La Conférence [...] décide que les remerciements qu'elle adresse au Conseil fédéral de la Suisse pour la gestion du Bureau International ainsi que l'expression de son désir de voir relever la position financière du Directeur et du Secrétaire de ce Bureau seront consignés au Procès-verbal». *Projet pour le Budget du Bureau international pour l'année 1872 adressée au Conseil fédéral,* 11 Juin 1872, in ITU-Corr, feuille n. 2/1, 1872.

Chapter 6
The *Bureau*cratisation of the Telegraph Union: St Petersburg (1875)

Introduction

St Petersburg was the last of the diplomatic conferences which led to the drawing up of a convention with the value of an international treaty, and for this reason needed to be signed by the diplomats representing the various member countries. As it was the last diplomatic conference, a more streamlined version was drawn up containing the Union's basic, unchangeable principles which were to remain in force until 1932, when the Madrid Conference decreed the birth of the International Telecommunication Union. Only technical delegations attended the conferences following St Petersburg, which were limited to producing documents concerning the technical and tariff regulations of the international telegraph service.

The three-year period (1872–1875) separating the Rome and St Petersburg Conferences was relatively peaceful, consolidating the new balance of power which had been forming over the previous years. By then, after the victories over Austria and France and the completion of its unification process, Germany had taken over the role of main political power on the continent and was on the way to a forceful industrialisation which would soon transform it into a leading economy. Another new and important actor, the States, was taking the scene by force. The Era of Reconstruction under Grant's presidency sealed up the bloody fractures left by the Civil War[1]. Re-found national unity and progressive control over

1 E. Foner and O. Mahoney, *America's Reconstruction: people and politics after the Civil War* (New York: HarperCollins, 1995).

the North American territory at the expense of the native populations[2] together with a progressive and intense industrialisation, all contributed to moving the focus away from Europe as centre of the world.

With the exception of Spain, torn apart by the Third Carlist War[3] (1872–1876), most European states were taken up with colonial expansion. Significantly, it was in this period when the annexation of India had been well completed and the majority packet in the Suez Canal acquired that the British Empire was officially born under the Disraeli government (1876)[4]. In general terms, this period started off a process of globalisation in international relations that lasted up to WWI, and remained unequalled until the end of the Cold War.

In the history of international communications the early 1870s saw the completion of the *Overland Telegraph Line* (1872) connecting Australia with the rest of the world[5] and the foundation in Bern of the Universal Postal Union (UPU), on the model of the Telegraph Union (1874)[6].

All of these events in international politics were bound to be mirrored in the balance of power underpinning the TU and were given full voice during the St Petersburg Conference. First of all, the new stability of European international relations, destined to last up to the outbreak of WWI, was reflected in the need to create stable bureaucratic structures and procedures able to guarantee the Union a continuity of action. Indeed, the St Petersburg Convention gave definitive approval to an organisation based on two principal elements, the two "gears", which had already been identified and named in the Rome Conference – the periodical meeting

2 A. Quinn, *Hell with the Fire Out: A History of the Modoc War* (Boston: Faber & Faber, 1997).

3 J. Extramiana, *Historia de las guerras carlistas* (San Sebastián: L. Haranburu, 1979–1980).

4 S. Mahajan, *British foreign policy, 1874–1914: the role of India* (London and New York: Routledge, 2002).

5 W. Lewis, S. Balderstone and J. Bowan, *Events That Shaped Australia* (Chatswood, N.S.W.: New Holland, 2006), 66.

6 G. A. Codding, *The Universal Postal Union: coordinator of the international mails* (New York: New York University Press, 1964); S. Fari, G. Balbi, G. Richeri, '"European Multilateralism' (1848–1865): a Telegraphic idea?" (paper presented at the 6th Plenary Conference of Tensions of Europe, Paris, September 19–21, 2013).

of plenary assemblies attended by delegates from all member nations and the continuous coordination of a permanent body. The St Petersburg Convention sealed the Bureau's central position in the functioning of the Union, so that Switzerland's hold over the administration became more and more permanent.

Secondly, Great Britain, which had won a major control of the global network thanks to its private submarine cable companies, pushed to obtain greater influence over conference decisions. This emerged in the continual interventions that the British, British-Indian and main submarine companies made, especially in the sittings dealing with tariffs. They were often coordinated to highlight the crucial British role in extra-European communications, in opposition to the consolidated powers of the preceding conferences, based on the Union's founding members. The expansion of European colonialism also affected the internal dynamics of the Union at St Petersburg. It played, in fact, a central role in the definition of a member state and attributing voting rights. The states with colonial possessions wanted to apply the rule of one vote per colonial administration (as had happened at Rome with Great Britain's and British India's separate votes). At the end, the nucleus of founding countries managed to impose the rule of one state one vote.

Finally the birth of UPU, which had taken place a year earlier, had a direct effect on the decisions taken by the assembly. Stephan, Minister of the German Posts and Telegraphs and main promoter of UPU was also delegate at St Petersburg and on more than one occasion managed to direct the decisions towards a greater convergence between postal and telegraph systems[7].

In a moment where great changes and the consolidation of organizational procedures and structures decided in the past lived side by side, Switzerland was able to tread softly and quietly and assert its role as guide to the Union.

---

7 Note for example the introduction at international level of the urgent/registered telegram and recorded delivery which copied similar postal services.

6.1 Continuity and change

As illustrated elsewhere, the Bureau carried out five important functions: collecting information about new technologies; publishing the *Journal Télégraphique*; sending official memorandums to member states; archiving documentation about changes in the norms; and offering consultancy for the national administrations about the technical features of the international service[8]. As recognised by most of the other member states, these functions were all carried out under the direction of the Swiss telegraph administration.

This guiding role was not determined as much by the fact the Bureau headquarters were in Switzerland but that the head answered directly to the Swiss Telegraph Administration and therefore indirectly to the Federal Council[9]. Besides acting as guarantor of the Bureau's efficiency, Switzerland was also called to prove in this field too that it was inspired by principles of neutrality and was not using its position of privilege to influence the TU's decisions to its own advantage. Opinions over how Switzerland stood differ. According to George A. Codding Jr., the Bureau treated procedural and administrative questions to exert the least possible influence over the general policies of the Union, or at least make it imperceptible to those who were not directly involved[10]. More recent studies however offer a different reading, and attribute to the Bureau a greater power than was thought, both in its technical consultancy with single administrations and with collecting proposals for modifying conventions and regulations[11]. In any case, in both circumstances the Bureau initiated correspondence with all contracting states and often added its own opinions to the answers, thus influencing the decisions of the single members[12].

---

8 Fari, *Una penisola in comunicazione*, 468.

9 See Chapter 4 and 5 on these topics.

10 Codding, *The International Telecommunication Union*, 51.

11 Fari, *Una penisola in comunicazione*, 469–475.

12 See for example the inquests about telegraph operators contracting lungs diseases and the establishment of an international telegraph school in ITU-Corr, dossier

To sum up, the period 1872 to 1875 confirmed the Bureau's role in regulating international telegraph relations despite the changes at the top. Chapter 5 above details how in 1869 the Federal Council had appointed the head of Swiss Telegraphs Lendi as Acting Director, and put off any definitive decision until the Rome Conference, while pointing at the same time to a viable solution. The delegates in Rome officially ratified the Federal Council's proposal in Lendi's absence on 24 May 1872. According to the *Journal Télégraphique*, Lendi's nomination was to be interpreted a duty to the person who "had successfully managed it in a temporary role"[13]. In spite of the apparently symbolic nature of the appointment, Lendi carried out his functions competently in the short time he was left to live. The documentation of the Post Department on the budgets of the Bureau[14] and International Telegraph School[15], show that he was active for a few more months.

Lendi died on 12 January 1873 and besides entrusting the everyday running to the Administrative Secretary of the Bureau, the Post Department provided at once to the definitive appointment of the new director. Differently from before, it was decided not to wait for the opinion of the next Conference because "the inconveniences that could arise for running the business and developing the service of a new provisory organisation" forced the Federal Council to make a permanent nomination[16]. Differently from the later custom, the head

n. 74, 1870 and dossier n. 132, 1872. In both cases the Bureau's opinion expressed in the letter opening the inquest, influencing implicitly many administrations.

13 «l'avait provisoirement géré avec succès». "Charles Lendi," *Journal Télégraphique*, 1/25 (1873), p. 192.

14 *Lettre écrite par Lendi adressée au Département des Postes*, 28 Mai 1872 (in ITU-Corr, feuille n. 1/2, 1872) and *Lettre écrite par Lendi adressée au Conseil Fédéral*, 11 Juin 1872 (in ITU-Corr, feuille n. 2/1, 1872).

15 *Lettre écrite par Ernesto D'Amico adressée au Bureau International*, 6 Juillet 1872 (in ITU-Corr, feuille n. 132/1, 1872) and *Lettre écrite par le Bureau international adressée à Ernesto D'Amico*, 24 Juillet 1872 (in ITU-Corr, feuille n. 132/3, 1872).

16 «les inconvénients qui pourraient résulter pour la gestion des affaires et le développement du service d'une nouvelle organisation provisoire» and «une nomination définitive». *Lettre écrite par le Département des Postes adressée à le Directeur de l'Administration des Télégraphes*, 7 Février 1873, in ITU-Corr, feuille n. 1/12, 1873.

of the Post Department was not nominated, for the Federal Council chose to recall into service Louis Curchod since he "had already been designated by the unanimous desire of all the states at meeting in Vienna and that he had consequently all the guarantees that this nomination would be favourably accepted by the various Administrations and would answer fully to the requested conditions"[17].

Given that for the first time the Federal Council was appointing the permanent Bureau head, it intended to select a person with guaranteed international prestige, whose performance was already highly rated by the TU delegates, and what was more important, Curchod had already been given this role by the Vienna Conference. Yet again, the Federal Council gives the impression of following or almost anticipating a desire then expressed by all the member states.

In spite of the changes – which in four years had brought to the top two directors and for a few months an administrative secretary – what emerges from the documents[18] about the period in which the Bureau had to face changes at the top is that it managed to keep a noteworthy stability. It was showing that it had reached such a level of bureaucratisation in the Weberian sense of the efficiency and rationality[19] that it was able to carry out its functions, whatever the circumstances. In spite of appearances, this rational management of international

---

17 «avait déjà été désigné par le veut unanime des délégués de tous les Etats réunis à Vienne et qu'il avait dès lors toutes les garanties désirables que cette nomination serait favorablement accueillie des diverses Administrations et répondrait pleinement aux conditions voulues». *Ibidem.*

18 As was recognized in the 1873, "The Federal Council had to provide again for the Direction of the International Bureau left suddenly vacant by the death of Mr Lendi. It called Mr Luois Curchod [...] This change in the person in charge of the service has not led to any modification in the conditions organizing the Bureau" («Le Conseil fédéral a dû pourvoir à nouveau à la Direction du Bureau International devenue subitement vacant par la mort de M. Lendi. Il a appelé M. Louis Curchod [...] Ce changement dans la personne placée à la tête du service n'a, d'ailleurs, amené aucune modification dans les conditions de l'organisation du Bureau»). *Rapport de gestion du Bureau international pour l'année 1874,* ITU-Corr, feuille n. 3/1, 1875.

19 See M. Weber, *Economy and society; an outline of interpretive sociology* (New York: Bedminster Press, 1968).

telegraphic relations was not in contradiction with the substantially proactive role of the Bureau.

6.2 Curchod mentoring the Russian delegate

Among the tasks given to the Bureau in the Rome Regulations was to collect proposals for changes/amendments to the articles in the convention and prepare the documents for the delegates at the coming conference (art. 34).

After a brief debate over which topics to assign to the Convention or the Regulations, the Rome Conference had decided to entrust the Bureau with a last, but very important task, i.e. prepare a: "*modification* of the Convention and Regulations which must be communicated to all the Offices about a year before the next Conference"[20]. In other words, the Director would be called on to work on the text of a proper constitution for the Union, based on the Rome version.

In reality, it is not clear from the Convention, Regulations or Minutes of the Rome Conference if the new convention project was to be masterminded or simply assembled by the Bureau. In a literal interpretation, the Bureau was to limit itself to collecting and publishing proposals coming from administrations belonging to the Union. Furthermore, following the custom of the previous conferences, the creative and proactive role in drawing up the Convention was to come from the administration organizing the conference, in this case the Russian one.

This ambiguity emerged and was resolved in a correspondence between the Bureau and the Russian administration over an apparently different topic: the voting system to adopt in the coming Conference. The exchange of letters also clarifies how important the Bureau and its

20 «*remaniement* de la Convention et du Règlement qui devrait être communiqué à tous les Offices environ une année avant la prochaine Conférence». *Documents de la conférence télégraphique internationale de Rome*, p. 443, [*our italics*].

director had become. It began with a letter sent by Lüders, the head of Russian Telegraphs, to the Director of the Bureau, bringing up a question which had been left unresolved in Rome, i.e. "if the vote should belong to the States in the diplomatic sense of the word or to the Telegraph Administrations"[21].

As organizer of the coming Conference, the Russian administration was to assume the chair and as such would have to decide if to allow the issue of the colonial vote to be aired afresh. On the invitation of the Italian administration, the Russian Director expressed his desire to settle the question for good:

> Our government, with its task of preparing the next reunion is also of the opinion that the system of voting, because of its importance, cannot be left in this state of indecision, and it takes as its special mission the task of regulating it as quickly as possible.[22]

Lüders then asked the Bureau to set in motion an opinion poll among other administrations in order to find out what they thought about the voting procedure, a task which was carried out by the secretary De St-Martial, acting head of the Bureau[23]. Most of the member administrations were against the multiple vote and those that were in favour like Spain and Italy hoped to gain from the geomorphologic, demographic

---

21 «le vote devait appartenir aux Etats dans le sens diplomatique du mot ou bien aux Administrations télégraphiques». *Lettre écrite par l'Administration Russe adressée au Bureau International*, 18 Décembre 1872, in ITU-Corr, feuille n. 29/1, 1872. In fact, as mentioned above in the introduction to the Rome Conference, Great Britain had brought one delegation representing its national networks and another speaking for its Indian dominions. As an exceptional measure and only after an explicit request from the British government, was it decided to give both delegations the right to vote, but only for the Rome sittings.

22 «Notre Gouvernement à qui il appartient de préparer la prochaine réunion est également d'avis que le système de votation, en raison de son importance, ne saurait rester dans cet état d'indécision; et il prend pour mission spéciale de la régler dans le plus bref délai». *Ibidem*.

23 *Lettre écrite par le Bureau International adressée à toutes les Administrations*, 10 Janvier 1873, in ITU-Corr, feuille n. 29/2, 1873.

and telegraphic advantages of their countries. Whatever, Curchod had in the meantime resumed the role of director again and took the project of the new convention firmly into his hands by means of the survey results and a dense correspondence.

In the summer of 1873, when the survey was coming to its end, Curchod decided to communicate the provisory results to the Head of Russian telegraphs in a somewhat surprising way, sending him two letters on the same day. The first had an official character and reported in a coldly professional way the opinions of all the administrations without ever quoting Curchod's[24]. The second was *confidentielle*, a kind of private communication between Curchod and Lüders[25], as the former was carefully to state[26]. More interesting than the first letter, the second can be read as Curchod's covert attempt to influence the preparatory stage of the conference. He was in fact moving along the line of a normative vacuum. He knew that the preceding documents did not attribute either to the Bureau or the Russian Administration the task of rewriting the new convention and therefore made a successful attempt to change the situation in his favour.

He declared very directly that the voting question was not in itself very important, almost as though to play down emphasis Lüders was giving it. Nevertheless, he wanted to furnish a reply, and taking the cue from Vinchent, his historical ally/adversary in various conferences, he declared his opposition to a multiple vote. He was in favour of keeping one vote/one state, and left open the question of giving the vote to multiple administrations within a state. Nevertheless, probably because he was reluctant to oppose Lüders and wanted show yet again his level of professionalism, he suggested at the end of his letter a simulation

24 *Lettre écrite par le Bureau International adressée à l'Administration Russe*, 11 Juillet 1873, in ITU-Corr, feuille n. 29/20, 1873.

25 *Lettre écrite par le Bureau International adressée à l'Administration Russe (confidentielle)*, 11 Juillet 1873, in ITU-Corr, feuille n. 29/22, 1873.

26 "it can only have, in any way [...] a personal character for me" («ne peut avoir, d'ailleurs [...] qu'un caractère tout personnel de ma part»). *Ibidem*.

of a multiple vote[27]. With in mind Spain, Italy and Russia's proposals, Curchod formulated a series of indicators (e.g. the number of inhabitants and territory size) with which to allocate the number of votes to the different states. The aim was obviously to give a scientific backing to his opinion, but in such a way as not to appear factious or presumptuous in Lüders' eyes.

The real reason for the confidential letter to Lüders was not so much to fix the voting question but to let the Russian know what Curchod thought about the future development of the Union. He began to go into the matter by commenting on the Italian's observations on the multiple vote. D'Amico had emphasized the fact that before discussing voting procedure they needed to know what type of document would be approved in the St Petersburg and successive conferences. D'Amico was in fact already drawing a distinction between diplomatic conferences with a one vote/one state rule and administrative ones where various administrations within the same state would all have the right to vote. Starting from this position Curchod illustrated his plan for St Petersburg and successive conferences:

> I think that the Paris Convention should be replaced at St Petersburg by *a new international treaty brought about via the diplomatic channels*. This treaty *would only contain the basic principles* which have been the rule since the Paris Conference and even before and which have never give rise to any serious debate. It will be decided upon and signed by all those states who would like to take part in the conditions which I feel must have to put aside giving voice more to some than to others; in the heart of the diplomatic conference there would be no real vote as such, a complete agreement is necessary but while respecting diplomatic usage, this diplomatic conference could perfectly well delegate to administrative conferences the job of formulating, pending government approval all kinds of measures (regulations and tariffs) needed to implement the conventions. In delegating this task it could without doubt organise administrative conferences and in particular the way of voting which will have to be respected and that will be able to, of common agreement, hold account in

27 The term simulation is used here in its technical-mathematical meaning, Curchod simulates the application of a new regulation by making a series of calculations which bring to an understanding of how a multiple vote would be divided among the various member states. He used mathematical simulation very much as happens nowadays.

the repartition of ways the number of distinct administrations and their relative importance [...] *To sum up on one side an unchangeable international treaty until notice of termination and on the other tariffs which lend themselves to necessary changes* and submitted to periodic reviews according to the forms dictated by the treaty itself.[28]

Thus for the first time Curchod explained how the organization of the Union was to be from St Petersburg onwards. He spoke for the first time of convening a last conference in which diplomats, the political representatives of national governments, would sign an unchangeable convention. For the first time he mentioned the fact that St Petersburg would be followed by administrative conferences alone.

While the Bureau had been given the task of drawing up a convention in Rome, the idea of making a permanent change to the nature of the conference was aired for the first time in this letter. Curchod had a very clear idea of the evolution of international conferences and he was equally aware that he could not impose it on Lüders and would have to

28 «Je pense que la Convention de Paris devrait être remplacée à St. Petersburg par *un nouveau traité international conclu par la voie et dans la forme diplomatiques*. Ce traité ne *contiendrait que les principes fondamentaux* qui ont fait règle depuis la conférence de Paris et même auparavant, et qui n'ont jamais donné lieu à aucune discussion sérieuse. Il serait arrêté et signé par tous les Etats qui voudraient y prendre part dans des conditions qui me paraissent devoir écarter toute idée de donner plus de voix aux uns qu'aux autres; dans le sein de la Conférence diplomatique il n'y aurait pas de vote proprement dit, il y faut un accord complet mais cette Conférence diplomatique, tout en respectant les usages de la diplomatie pourrait parfaitement déléguer a des conférences administratives le soin d'arrêter, sous réserve de l'approbations des Gouvernements respectifs, les mesures de tous genres (règlement et tarif) nécessaires à l'exécution de la Convention, et en délégant ce soin elle pourrait aussi, sans aucun doute, organiser des Conférences administratives et en particulier le mode de votations qui devra y être observé et qui pourra, d'un commun accord, tenir équitablement compte, dans la répartitions des voix du nombre des administrations distinctes et de leur importance relative. [...] *En résumé, d'une part un traité international immuable jusqu'à dénonciation d'autre part les règlements et les tarifs se prêtent sans cesse aux changements nécessaires* et soumises à des révisions périodiques selon les formes dictées par le traité lui-même». *Ibidem*, [*our italics*].

negotiate. The following passage clarifies the point and highlights his diplomatic skills:

> I think I must communicate to you with this attachment the convention project I have drawn up, to put in order my ideas, in applying preceding principles and taking as my base the Paris Convention revised at Rome. In my eyes this project is what should have appeared in the diplomatic convention, if it had been concluded at Rome, having found its place in the regulations and tariffs. I do not for the moment insist in length on this project; I have said enough so that *you can* [...] *form an opinion and I beg you let me know as soon as possible.* The work on certification prescribed by the Rome Conference can only be usefully undertaken when you have drawn up the plan. In effect it is on your Department that the direction of operations for the St Petersburg Conference falls, and *I cannot and must not be more than an instrument in your hands* for the preparatory work.[29]

Curchod was of course well aware he was no simple instrument in Lüders' hands. Quite the reverse, for he attached to the above declaration a project for the convention which not only Lüders would make totally his, but would be discussed and passed without any opposition in St Petersburg. Several studies, in fact, agree on the fact that the Bureau in the person of Curchod reached the apogee in its influence over the procedures and contents to be approved by the Union in preparing the documents for St Petersburg[30].

29 «Je crois devoir *vous communiqué* ci-joint le projet de *convention que j'ai préparé*, pour me fixer les idées, en application des principes qui précédent et en prenant pour base la convention de Paris révisée à Rome. Ce projet représente donc, à mes yeux ce qui aurait dû figurer dans la convention diplomatique, si on en avait conclu une à Rome, tout le reste ayant trouvé sa place dans le règlement et dans le tarif. Je n'insiste pas pour le moment plus longuement sur ce projet; j'en ci dis assez pour que *vous puissiez* [...] *former une opinion et je vous prie de me la faire connaître le plus tôt possibles*. Le travail de certification prescrit par la Conférence de Rome ne pourra être utilement entrepris que lorsque vous aurez arrêté le plan. En effet, c'est à votre Département qu'incombe incontestablement la direction des opérations de la Conférence de St. Petersburg et *je ne puis ni ne dois être*, dans les travaux préparatoires qui la concernent, *qu'un instrument entre vos mains*». *Ibidem*, [*our italics*].

30 Codding, *The International Telecommunication Union*, 51; Fari, *Una penisola in comunicazione*, 473.

Though Lüders' answer referred explicitly to Curchod's official letter[31], his words show he must have read the confidential one carefully. In fact, he asked Curchod to explain how to cast his vote in the administrative and diplomatic conferences:

> I beg you, dear Director, to submit again this object to the examination of the Administrations interested and collect their opinions on the decision if the vote should belong to the States in the diplomatic sense of the word or to the administrations and then give me the results together with your opinion on the subject.[32]

It is clear that once he had understood why Curchod judged the multiple vote pointless, Lüders wanted a proposal on how to solve the problem of the colonial vote. He therefore asked Curchod to open a new consultation and then give him the results together with his own opinion. The two heads met in Vienna most probably in August/September 1873[33] and there Curchod showed Lüders his project for amending the voting procedures: 1) one vote per state for approving the last diplomatic convention; 2) possibility to allocate a vote to various administrations

---

31 *Lettre écrite par le Bureau International adressée à l'Administration Russe*, 11 Juillet 1873, in ITU-Corr, feuille n. 29/20, 1873.

32 Je vous pries, Monsieur le Directeur, de vouloir bien soumettre de nouveau cet objet à l'examen des Administrations intéressées et en recueillir les opinions sur la décision si le vote devait appartenir aux Etats dans le sens diplomatique du mot ou bien aux administrations et ensuite me communiquer le résultat simultanément avec votre propre appréciation sur la matière. *Lettre écrite par l'Administration Russe adressée au Bureau International*, 27 Août 1873, in ITU-Corr, feuille n. 29/23, 1873, *underlining in the original.*

33 "In conformity with what has been recently agreed among us at Vienna, I have the honour to address you the attachment, after having revved the draft and the voting proposal that I wrote with your suggestions and approval. It is a simple outline rather than a complete profect, even suppposing that its contents to be approved and adopted wholly by you" («Conformément à ce qui a été convenu récemment entre nous à Vienne, j'ai l'honneur de vous adresser ci-joint, après en avoir revu la rédaction, le projet de vote que j'ai écrit sous votre inspiration et avec votre approbation. Il s'agit plutôt d'une simple esquisse que d'un projet complet, car à supposer même que son contenu fut approuvé et adopté en entier par vous»). *Lettre écrite par le Bureau International adressée à l'Administration Russe (confidentielle)*, 28 Septembre 1873, in ITU-Corr, feuille n. 29/25, 1873.

belonging to the same state providing the government in question made an explicit request and the delegates of each administration were present at the conference. An official letter from Curchod to Lüders of 24 October marked the final closure of the inquest[34]. Curchod had not only greatly influenced the final version of the voting procedure norm, so crucial for functioning of the conferences, but he had set up the future agenda of the Union and its aim to become a stable structure destined to last in time. What is noticeable is that he deployed the traditional bureaucratic and administrative instruments the conferences had given him to reach these aims. He never formally passed the limits set by convention rules and regulations and, in observing them, managed to pilot almost all the most important decisions. Coherent, he behaved in exactly the same way in St Petersburg.

6.3 The Bureau-cratic system

On 1 June 1875 the Fourth Conference of the TU opened in St Petersburg. The delegates realized at once that the assembly would not be introducing new norms as much as dealing with establishing immutable rules to give force and continuity to the Union's organizational structures. The Russian minister opening the conference, declaring it

> would have the great honour of *establishing permanently* in a kind of succinct international Code those of the rules created to regulated universal telegraph relations and *established by you in previous conferences*, which ten years of study and usage have above all shown the efficacy and urgency.[35]

34 *Lettre écrite par le Bureau International adressée à l'Administration Russe*, 24 Octobre 1873, in ITU-Corr, feuille n. 29/28, 1873.

35 «aura l'insigne honneur *d'établir définitivement* dans une espèce de Code international succinct celles, parmi les règles appelées à présider aux rapports télégraphiques universels et *établies par vous dans les précédentes Conférences*, dont dix années d'études et d'usage ont surtout démontré l'efficacité et l'urgence». *Documents de la Conférence Télégraphique Internationale de St Petersburg*, Bureau International des Administrations Télégraphiques, Berne 1875, p. 268, [*our italics*].

The head of the Russian Telegraphs was more explicit, as was the Chairman Lüders:

> The Rome Conference unanimously expressed the desire to change the convention into a less extensive and complicated act and tasked the International Bureau with drawing up the elements. This work is completed toady. It is for us to now examine well the pieces which have been sent to you and all have in their hands. It is a question of looking for a solution to the regulation questions which will improve the present convention and find the way to make a *very simple, practical, general convention, of a nature to facilitate adhesion.* In effect, we must facilitate it and make it as desirable as possible, so that our Telegraph Union, which is already a big family, grows even bigger and spreads over a wider territory.[36]

Figure 17: The technical delegates at the St Petersburg Conference, 1875.
Source: ITU Archives.

---

36 «La Conférence de Rome a exprimé à l'unanimité le désir de changer la Convention en un acte moins étendu et moins compliqué et elle a chargé le Bureau international d'en préparer les éléments. Ce travail est fait aujourd'hui. Il s'agit pour nous maintenant de bien examiner les pièces qui vous ont été remises et que vous avez tous entre les mains. Il s'agit de chercher, pour les questions règlementaires, une solution, qui améliore la convention actuelle et de trouver le moyen de faire une *convention bien simple, bien pratique, bien générale, de nature à faciliter l'adhésion*, en effet, nous devons la faciliter et la rendre *désirable autant que possible*, pour que notre Union télégraphique, qui est déjà une grande famille, s'agrandisse encore et s'étende sur un domaine plus vaste». *Ibidem*, p. 271, [*our italics*].

While in the opening of the proceedings Lüders pointed to the Bureau's crucial role in preparing the documents, consecration came at the end, in the closing sitting:

> It was a very complicated task which the Conference was not able to take over in the course of the works, but it has been so well prepared by the International Bureau that the Conference had very little to change [...] These considerable improvements have been obtained without causing great work for the Conference and especially the Commissions.[37]

What had happened was that the most important norms had already been decided and inserted by the Bureau in the project of the convention and regulations drawn up before the opening. All the assembly and commissions did was to make some "considerable changes" without altering the contents.

Probably in order to tackle the needs concerning the drawing up a definitive convention, the St Petersburg Conference differed in both the preparatory stage and the assembly. During the first sitting, Curchod explained carefully what had happened in the phase of documentation preparation:

> Mr Curchod [...] announces that he has submitted first of all to the Russian Administration, as tasked by the present Conference organisation a convention pre-project, which has been sent to all the states, as an attachment to the Imperial Government's circular.
>
> *This first project, in which the International Bureau in line with the Russian Administration's idea, only included the essential dispositions,* consecrated in some way by a continuous experience and general adhesion, has undergone, following the codification of the regulations, some further modifications which do not distance from the principles adopted for its elaboration. The result of this double work were sent to all Administrations in January 1875, to be studied before the Conference. Mr Curchod notes about this that the examination will show that

37 «C'était une œuvre très compliquée dont la Conférence n'aurait pas pu se charger dans le cours de ses travaux, mais elle a été *si bien préparée par le Bureau international que la Conférence n'a eu que peu de changements á y apporter*. [...] Ces améliorations considérables n'ont pu s'obtenir sans causer de grands travaux à la Conférence et surtout aux Commissions». *Ibidem*, p. 648.

the Paris Convention dispositions, as coming out of the Rome deliberations, have been reproduced in the new convention and regulation projects. As for the propositions of the different Offices, they have only been produced in following the old text and the International Bureau had to gather and co-ordinate them in two different book, one of which, the yellow book follows the order of the subjects in the Rome text and presents them exactly as the Administrations themselves had produced them. The other, blue book classes them in the order of the new projects, only adding formal modifications necessary in re-touching the original plan. Reference numbers create a concordance between the two books.[38]

From a rapid comparison between the acts of the St Petersburg Conference and the previous ones, it emerges that the discussions over various convention and regulation articles were decidedly shorter. Most of the norms were read and approved without giving rise to the heated debates that had occurred in the past. It was generally the result of a rigorous procedure conducted by the Bureau, which organized a series of preliminary phases designed to obtain the delegates' tacit

38 «M. Curchod [...] fait connaître qu'il a soumis, d'abord, à l'Administration russe, comme chargée de l'organisation de la Conférence actuelle, un avant-projet de Convention qui a été communiqué à tous les Etats, en annexe à la circulaire du Gouvernement impérial. Ce *premier projet, où le Bureau international, d'accord avec les idées de l'Administration russe, n'avait compris que les dispositions essentielles*, consacrées, en quelque sorte, par une expérience constante et une adhésion générale, a subi, à la suite de la codification du Règlement, quelques modifications ultérieures qui ne s'écartent point, d'ailleurs, des principes adoptés pour son élaboration. Le résultat de ce *double travail* a été transmis, au mois de janvier 1875, à toutes les Administrations, pour pouvoir être étudié avant la Conférence. M. Curchod fait observer à ce sujet que cet examen permet de constater que toutes les dispositions de la Convention de Paris, telle qu'elle est sortie des délibérations de Rome, ont été reproduites dans les nouveaux projets de Convention et de Règlement. Quant aux propositions des différent Offices, ceux-ci n'ont pu les produire qu'en suivant l'ancien texte et le Bureau international a dû, dès-lors, les réunir et les coordonner dans deux cahiers différents, dont l'un, le cahier jaune, suivant l'ordre des matières du texte de Rome, les présente exactement comme les Administrations elles-mêmes les avaient produites, et dont l'autre, le cahier bleu, les classe suivant l'ordre des nouveaux projets, en apportant seulement les modifications de formes nécessitées par le remaniement du plan primitif. Des chiffres de renvoi établissent la concordance entre l'un et l'autre de ces cahiers». *Ibidem*, p. 285, [*our italics*].

agreement, and lead to sittings where there was no need for debating or diverging. The Bureau had drawn up the convention and regulation project, submitted it to the administrations and received their comments. In the first sitting Curchod presented the project together with the amendments from the various delegations and suggested a procedure for approving the acts which with some minor changes was accepted. The procedure provided for: 1) an initial reading of the project for the Convention, article by article 2) direct approval of the articles or re-reading of the regulation in the case of disagreement; 3) reading the regulation articles, one by one; 4) in the case of agreement, direct approval of the articles of the regulations and the convention norms pending from phase 2; 5) in the case of disagreement over an article, deferment to the relative committee[39]; 6) committees formulating proposals for new articles; 7) after the initial reading of the convention and relations, presentation of committee's opinion of articles pending on phase 5 (approval/rejection of changes); 8) in the case of the committee's proposal being judged favourably by the delegates, delegates' direct approval of the article; 9) in the case of the committee's proposal being judged negatively, opening of a discussion between the different positions and in the case of no agreement, article referred to Bureau; 10) collection of divergent opinions, Bureau formulating compromise; 11) almost automatic approval of the article formulated by the Bureau.

All this complex bureaucratic procedure concentrated much power in the Bureau's hands. It had drawn up the convention/regulation project, collected the relative amendments and established the procedure for debating the articles, as well as acquired the power to

39 The committees were bodies already formed in the Rome Conference to speed up discussion and approval proceedings of convention and regulation articles. While however in Rome they were informal bodies with a prevalently consulting power, in St Petersburg their function was to formalize. Two committees were named during the first sitting, the first for tariff matters while the second focused on doubts about regulation articles. In St Petersburg the committees acted with the power of parliamentary ones and gave their opinion on articles which had aroused discord among the delegates. A committee's opinion was normally accepted by the assembly.

close the approval procedure of the norms in the case of disagreement among the delegates. If an article was not approved during the sittings, phase 10 actually allowed the Bureau to formulate a compromise article keeping account of the needs of all. This procedure was really formalising a custom that had been shaped back during the conferences of Paris, Vienna and Rome. Faced by the considerable discord among delegates in Paris and Vienna, Curchod had often successfully intervened and suggested a compromise article to please all involved. Given Lendi's absence and Curchod's late entry as Swiss representative, Vinchent had covered the same role in Rome. The formalisation of the procedure implied two new important consequences. The first was the formal recognition of the Bureau's intermediary role also within the conferences. The second was the Bureau drafting definitive convention articles which delegates had not agreed over, which were therefore the most delicate issues (like the antidumping regulation).

It goes without saying that such a high-profile role would not have been assigned to the Bureau if the delegates had not trusted it entirely and been convinced of the high level of competency and impartiality of its director. Their esteem transpires from their decisions and declarations recorded in the Acts of the Conference. A first example is to be found in the reading and approval of article 13 of the convention, which foresaw a change of regulation in any moment providing all the administrations involved were in agreement. As a permanent body, the Bureau had to collect the opinions and in the last resort ascertain the presence of unanimity. In spite of the opposition of a few delegates, the conference approved the articles which had been drawn up, with the knowledge they were also giving the Bureau a considerable power:

> M. Vinchent notes that Russia's proposition would have the only aim of suppressing the faculty to modify regulations in the intervals between two conferences. This faulty is however in his eyes very useful and is *one of the most interesting attributions of the International Bureau.* Perhaps for the greatest number of propositions and the most important ones, agreement will not be reached; but the study carried out by the International Bureau, does not beat less fruits for this, for the

question arrives mature at the Conferences and allows them to make a more rapid and enlightened solutions.[40]

Further proof of the esteem enjoyed by Curchod and the Bureau's performance came out during the discussion of the articles concerning the Bureau itself. The Italian delegate asked for the possibility to express a personal opinion before the vote on the articles concerning the Bureau, and asked for Curchod and his secretary to leave the hall. D'Amico then waxed eloquent:

> Mr D'Amico proposes first to the Conference to express to the Swiss Federal Government the recognition of all the Administrations for its valiant and useful intervention in the organisation of the International Bureau and *the services rendered by this body to European telegraphy.* He proposes the same to manifest to the International Bureau's head and secretary entire satisfaction for their efforts and success in the intelligent and economic management of the bureau.[41]

Words turned into deeds when the same D'Amico made a proposal to increase the funds allocated to the Bureau, which won a unanimous vote. It was not of course a direct salary increase for Curchod, given that only the Swiss government could decree that. Nevertheless, such a generous rise in the resources at his disposal (from 50,000 to 60,000 francs) was a very precise invitation to the Federal Council to raise salaries for Curchod and his secretary.

---

40 «M. Vinchent constate que la proposition de la Russie ne saurait avoir objet que de supprimer la faculté de modifier le Règlement dans l'intervalle de deux Conférences. Cette faculté est cependant très-utile à ses yeux, et forme *une des attributions les plus intéressantes du Bureau international.* Il est possible que pour le plus grand nombre et pour les plus importantes des propositions, l'accord ne s'établisse pas; mais l'étude provoquée par le Bureau international n'en porte pas moins ses fruits, car la question arrive de la sorte mûrie aux Conférences et leur permet de prendre une solution plus rapide et plus éclairée». *Documents de la Conférence Télégraphique Internationale de St Petersburg*, p. 303.

41 «D'Amico propose, d'abord à la Conférence d'exprimer au Gouvernement fédéral suisse la reconnaissance de toutes les Administrations, pour sa haute et utile intervention dans l'organisation du Bureau International et *pour les services rendus par cette institution à la télégraphie européenne.* Il propose également de témoigner au directeur et au secrétaire du Bureau international la satisfaction complète que méritent leurs efforts et leurs succès dans la gestion intelligente et économique de ce bureau». *Ibidem*, p. 544, [*our italics*].

On 19 July 1875, the St Petersburg Conference closed and for the last time the new convention was signed by the diplomats representing the member states. The Bureau had scrupulously overseen all the operations of drawing up, discussing and approving the new convention. It came away from St Petersburg notably reinforced in prestige, competences and functions. Its role in these terms was destined to last in time, since it was defined very precisely by the Convention, which would undergo no changes for the next sixty years.

6.4 The separation of the ways

At the Paris and Vienna Conferences Curchod, then the Swiss delegate, had adopted what was often defined as the "reciprocal concessions" strategy, as he sought to harmonize his functions as intermediary between the various administrations and at the same time defender of national interests. In this way he had obtained the important result of establishing the Bureau, the permanent body placed directly responsible to the Swiss Post Department (and then the Federal Council). As we have already seen, in Rome Lendi was to have represented the Bureau as its head and the Federal Council as its delegate. Since his illness stopped him from taking part, his secretary De St-Martial stepped in to represent the Bureau while Curchod, already at the conference as delegate of a French submarine cable company took over as Swiss delegate. Already in Rome, Curchod stood out as the most likely candidate for running the Bureau and therefore had to move with great diplomatic caution, showing, as he always had in the past, a great ability in creating compromise among the delegates.

In St Petersburg, the conditions were very different right from the start. While Curchod spoke for the Bureau, the Swiss Federation he had represented in the previous three conferences now had in Bernard Hammer (Swiss ambassador in Berlin) and August Frey (head of Swiss Telegraphs) both an administrative and diplomatic representative.

Figure 18: August Frey, head of Swiss Telegraphs from 1872 to 1889 and head of the Bureau for few months during 1890.

Source: PTT Archives.

For the first time Switzerland could count on two delegates mandated to act alone, while it lay to Curchod as director of the Bureau to continue mediating and following Switzerland's aims as before, and the delegation would act only in the nation's interests to obtain as many advantages as possible. On several occasions during the proceedings, Hammer clarified how he saw the role of the delegate:

> The concept of the Conference as a reunion of experts is, furthermore, in the thought of Mr Hammer an incomplete concept, for there are the delegated, and he foremost who while not being experts from a telegraphic viewpoint do not have fewer qualities for defending and upholding the interests of their countries in their words and votes.[42]

42 «La conception de la Conférence comme une réunion d'experts, est, d'ailleurs, dans la pensée de M. Hammer, une conception incomplète, car il y a des délégués, et lui-même tout le premier, qui, tout en n'étant pas des experts au point de vue

> As for the concept of the Conference as a reunion of experts, it is not wholly just, for the delegates do not only bring the contribution of their understanding but they also have the defence of the interests they represent.[43]

The Swiss delegates received instructions from the Federal Council which were no different in scope from those received by other delegations. They were substantially told to follow the general line of reform for the convention suggested by the Bureau except for particularly delicate questions. The principles which had always inspired Swiss policies over telecommunications had remained substantially the same. The Federal Council instructed its delegates to reject the proposals put forward by Austria-Hungary and British India calling for a reform which would limit guarantees over delays and sudden hikes in tariffs both in Europe and the rest of the world[44]. As for the tariff policy, the Swiss instructions followed the same line as other countries and were directed at preserving national interests, which in the Swiss case meant opposing any measure attempting to give particular favour to the non-European administrations[45].

For this reason the Swiss delegates only resorted to strong action about specific norms and never damaged the institutional and bureaucratic organization created by Curchod. Swiss dissent surfaced particularly over two issues, which were the proposed introduction of urgent telegrams and changes to tariff conditions in Europe. Both in Vienna and Rome the Italian delegate D'Amico had tried to introduce the concept of urgent telegrams having precedence over ordinary ones but at a higher cost. However, his proposal had only received the support of Belgium, which like Italy had already successfully introduced

---

télégraphique, n'en ont pas moins qualité pour défendre et soutenir les intérêts de leur pays par leur parole et par leurs votes». *Ibidem*.

43 «Quant à la conception de la Conférence comme une réunion d'experts, elle n'est point complètement juste, car les délégués n'apportent pas seulement le concours de leurs lumières, mais ils ont aussi la défense des intérêts qu'ils représente». *Ibidem*, p. 512.

44 *Ibidem*.

45 *Ibidem*.

urgent telegrams into its national service. Yet at St Petersburg the Italian proposal met the favour of all the administrations with the exception of Switzerland and Great Britain. Switzerland's opposition was no new turn, but Hammer's stance in comparison to Curchod's in previous conferences appears far more ideological, clearly reflecting the radical nationalist positions of the Swiss government. This was the first time since the creation of international telegraph bodies that Switzerland had decided to separate domestic and international communications policies.

> M. Hammer has to declare that the Federal Council's instructions enjoin him to oppose the adoption of urgent telegram. *These instructions rest on the reason that an exception made in the favour of wealth is odious in a country that has a strong feeling of equality.* On the other side he would like to know the guarantee there would be for ordinary telegrams to be transmitted in a given time when precedence is taken by privileged correspondence. If the order of transmission could be taken as a rule, and the importance of content instead of granting priority to those who can afford higher charges, urgency could be justified logically. But a similar valuation is impossible, an in application, *priority would be reduced to a question of money.* Now the thing presents this dual danger, that the public sees it a fiscal measure by the administrations to make a greater profit with no increase in work and on the other that telegraph personnel would be inclined to neglect ordinary correspondence and recommend urgent transit to senders. These inconveniences would be greater with the rise in the number of urgent telegrams and if, on the contrary, the number is insignificant, the measure would present neither the interest or utility its supporters find in it.[46]

46 «M. Hammer doit déclarer que les instructions du Gouvernement fédéral lui enjoignent de s'opposer à l'adoption des dépêches urgentes. *Ces instructions s'appuient sur ce motif qu'une exception faite en faveur de la richesse présenterait quelque chose d'odieux dans un pays qui a un profond sentiment de l'égalité.* D'une autre côté, il voudrait savoir quelle garantie il y aurait pour que les dépêches ordinaires fussent transmises dans un temps déterminé, quand leur transmission viendrait à être primée par des correspondances privilégiées. Si l'on pouvait prendre pour règle de l'ordre de la transmission, l'importance du contenu des dépêches, au lieu d'accorder la priorité à l'expéditeur qui peut payer une plus forte taxe, l'urgence se justifierait logiquement. Mais une pareille appréciation est impossible, et dans l'application, la *priorité se réduit à une question d'argent.* Or, la chose présente, dès lors, ce double danger, d'une part, que le public y voie

In other words, the official reason for Swiss opposition lay in the presumed discriminating nature of urgent telegrams, which were accused of guaranteeing privileges and limiting access to the citizenry. A more hidden reason was instead the fear that introducing urgent telegrams in a country dedicated to business and international commerce like Switzerland might send telegram tariffs right up (since everybody would want to use the priority service) and this would give the Swiss administration extra work, both as regards personnel and structures. In that period liberal administrators (and therefore the Swiss ones too) were convinced that urgent telegrams should be marginally dearer than the others, i.e. that the cost/price ratio would increase the overall outgoings for personnel and structures. It emerges from the comments of the Federal Council about both urgent telegrams and notifications, which Switzerland was against for the same reason:

> The Federal Council has already had the occasion to pronounce itself on the admission of this nature for the relations between Switzerland and abroad. It was principally guided by the consideration that it would never be admissible to offer an advantage to the public in the international service, though only apparent, without introducing it equally in the domestic service, but for the latter we could wait, in view of the natures of the correspondence, until at least a half is presented in the form of notification, and there is a reduction of about 200,000 francs receipts per year, without an appreciable compensation in the reduction of work. On the other hand, we believe we can admit notifications in the transit service, on one side because the existence of direct lines greatly reduce the work of the Swiss staff and on the other to keep on our lines transit which could be sent along other lines [...] In agreement with a decision previously taken by the Chambers and considering that allowing urgent telegrams can only have the consequence of depreciating and casting into neglect ordinary telegrams, we have told our delegation to reject the proposition about it.[47]

---

une mesure fiscale de la part des administrations pour retirer un plus grand profit sans accroissement de travail et, de l'autre, que le personnel télégraphique ne soit enclin à négliger la correspondance ordinaire pour recommander l'urgence aux expéditeurs. Ces inconvénients seraient d'autant plus sensibles que le nombre des dépêches urgentes s'augmenterait et si, au contraire, le nombre en était insignifiant, la mesure ne présenterait ni l'intérêt ni l'utilité qu'y trouvent ses partisans». *Ibidem*, p. 313, [*our italics*].

47 «Le Conseil fédéral a déjà eu l'occasion de se prononcer sur l'admission de cette nature pour les relations entre la Suisse et l'étranger, il a été guidé principalement

Both Great Britain and Switzerland adopted a hard line and were strangely little inclined to come to a diplomatic resolution. When the majority declared they were in favour of introducing urgent telegrams, Great Britain and Switzerland not only went into opposition, but declared that since the proposal was against their governments' mandates, they were ready to exercise their right to veto. This meant that because there was no unanimity, the introduction of urgent telegrams would not be written into the convention.

Following the wishes of the majority, the article on urgent telegrams was prepared to be inserted in the Convention[48]. Great Britain kept to its hard line and threatened that if the article came into force, it would not even allow urgent telegrams to transit through its territories. Switzerland agreed to allow transit, but not to urgent telegrams being sent from its territory[49]. The Swiss delegates probably took a step backwards on instructions of their government, in order not to hinder the diplomatic work of Curchod as Bureau head. In fact Curchod managed to propose a compromise article on urgent and registered telegrams, and telegraph notifications which could be accepted and voted by Great

par la considération qu'il ne serait guère admissible d'offrir au public dans le service international un avantage, quoique apparent seulement, sans l'introduire également dans le service interne, mais que pour ce dernier on pourrait s'attendre, en vue de la nature des correspondances, à ce que au moins la moitié serait présentée sous forme d'avis et qu'il en résulterait une diminution de recettes d'environ 200.000 Fr. par année, sans compensation appréciable par la diminution du travail. Par contre nous croyions pouvoir admettre les avis dans le service de transit, d'un côté parce que l'existence de lignes directes réduit considérablement le travail des employés suisses, de l'autre côté et notamment pour conserver à nos lignes le transit qui sans cela pourrait être détourné d'autres voies [...] D'accord avec une décision prise antérieurement par les Chambres et en considérant que l'admission de dépêches privilégiées ne peut avoir pour conséquence que de déprécier et de faire négliger les dépêches ordinaires, nous avions chargé notre délégation de rejeter la proposition y relative». *Message du Conseil fédéral à la haute Assemblée fédérale concernant la Convention télégraphique international de St-Pétersbourg*, 29 Novembre 1875, p. 10, in SFA, E52, 519.

48 *Documents de la Conférence Télégraphique Internationale de St Petersburg*, p. 438.

49 *Ibidem*, p. 446.

Britain and Switzerland, too[50]. So, while Hammer and Frey differed from Great Britain in backpedalling and accepting the transit of urgent telegrams, Curchod drafted an article which would allow a facultative urgent telegram service, thus meeting halfway Switzerland and Great Britain adamant in their refusal to allow an urgent telegraph service on their territory.

The Swiss also defended their national interests against a proposal to modify international tariff conditions. Article 15 of the new regulations fixed the number of words for a minimum telegram and the way of pricing the extra ones (by groups or single words). Surprisingly, the commission examining the various amendments came out for a revolutionary change which foresaw two movements: reduce the simple telegram from twenty to ten words and price any extra words individually. The new rules would be applied only to European telegrams since the rest of the world already enjoyed special conditions, with a 10 word count for a simple telegram and extra words costed separately. The commission suggested applying the same tariff system to European telegrams, but reactions were so strong that Vinchent even called for "the voters to be named, so as to be able to determine the influence each Office could have on their representative's decision"[51].

The discussion polarized around two groups. Those who upheld the motion included Italy, Germany and Austria, while those who spoke against it were Belgium, Switzerland and Denmark. The delegates in favour aimed at obtaining a general reduction in the tariffs by reducing the telegraph operators' workload thanks to the 50% cut in telegram length. They thought the tariff cuts would greatly increase international correspondence, and this would make up for the reduced income coming from lowered tariffs. Those against the proposal felt that with the different natures of international and inland correspondence, lowered tariffs would not lead to an increase in traffic. Thus, given that many nations (including Switzerland) had an inland service perennially running at a

50 *Ibidem*, pp. 631–639.

51 «connaître les noms des votants, afin d'apprécier l'influence que les intérêts de chaque Office pourraient avoir eue sur la détermination de leur représentant». *Ibidem*, p. 377.

loss, compensated for by income from international correspondence, a drop in the latter would have disastrous consequences on the telegraph administrations of a good part of Europe. This outlook terrified the majority of delegates, who ended by voting against the Commission's proposal. The simple telegram stayed at 20 words, with a 5 and not 10 word progression. Frey stood out as one of the great opponents to the reform:

> M. Frey declares himself equally for maintaining the present system and it is in this sense that he has voted in the Commission. The main reason is that he dreads the financial consequences of any charge reduction [...] The same scale would be applied to domestic and international correspondence. In the former any reduction easily excites the interests of families, who are very sensitive to a more contained charge; in the latter commercial telegrams, indifferent to price, continue to be exchanged.[52]

Vinchent was even more explicit:

> Telegraph administrations are almost all in the same situation of suffering losses for domestic correspondence which they compensate for entirely or partially by the benefits of international or transit correspondence, so much so that the cost price of domestic telegrams is higher than that of international, which in its turn passes that of transit.[53]

The opposition between those for and against tariff reform concealed a far more important contrast between two diametrically different

52 «M. Frey se prononce également pour le maintien du système actuel et c'est dans ce sens qu'il a voté au sein de la Commission. La raison principale est qu'il redoute les conséquences financières de toute réduction de taxe. [...] La même échelle ne saurait donc s'appliquer à la correspondance intérieure et à la correspondance internationale. Dans la première, toute réduction surexcite les intérêts de famille qui sont très sensibles à une taxe plus modérée ; dans la seconde, il ne continue à s'échanger que des dépêches commerciales, indifférentes au prix». *Ibidem*, pp. 374–375.

53 «Les administrations télégraphiques sont presque toutes dans cette situation de subir pour la correspondance intérieure des déficits qu'elles compensent, en tout ou en partie, par les bénéfices de la correspondance internationale ou de transit, tandis que le prix de revient de la dépêche intérieure, au contraire, par cela même qu'elle nécessite plus d'opérations, est plus fort que celui de la dépêche internationale, lequel dépasse à son tour celui de la dépêche de transit». *Ibidem*, pp. 376–377.

concepts of the international telegraph service. While all delegates appeared to agree that the domestic telegraph service had to be a universal service, with low tariffs favouring access to as many clients as possible, their opinions differed over international telegraphing. Opponents of the reform like Belgium and Switzerland held that the international service should target above all professionals (businessmen and news agencies) and felt therefore that lowering the tariffs would help not the general public but only economic operators who would be using the service anyway. For them the service was to be elitist and not universal, so that cutting tariffs was an illogical move.

It may seem strange that Switzerland took such a stand when in two of the three preceding conferences (Paris and Vienna) the government mandate had been to go on lowering as far as possible international tariffs. It was not however either a contradiction or even less a change in tactics. Switzerland and Belgium had always defended their position as transit nations and had managed to run their domestic services with very low tariffs only because of the huge income from international telegrams, mostly transit. In the first conferences, they had defended knocking down tariffs because they were so high in other countries that they hampered correspondence. The situation was reversed at St Petersburg, when the other states suggested a tariff reduction that would cut income and put the running of the domestic service into difficulty. Switzerland and Belgium therefore kept the same priority: they had to defend income from transit telegrams in order to maintain a domestic telegraph service accessible to all and sundry.

Both the opposition to urgent telegrams and tariff reform were actually marginal episodes in a conference where few rules were changed and, overall, the result was fully acceptable for the Swiss government:

> As it results from the above explanations of the service regulations contained in the attachment to the St Petersburg International Convention, besides some not very essential restrictions, several quite important improvements, and in the case of charges there is nothing that can harm the interests of the Swiss public.[54]

54 «Comme il résulte des explications ci-dessus que le règlement de service annexé à la Convention télégraphique internationale de St. Petersburg contient, à côté de quelques restrictions peu essentielles, plusieurs améliorations assez importantes

6.5 The Bureau as a Swiss body

The years between the Rome and St Petersburg conferences were essential for fixing the regulations of international telecommunications, not so much for the contents of the two conferences themselves as for the increasingly important role taken on by the Bureau.

While in the beginning the Bureau had been presented as a simple office for settling everyday matters, it had turned out to be the real driving force powering the development of telegraph relations. Besides carrying out its original functions efficiently, it also took over, thanks to Curchod, a proactive role which had not been foreseen when it had been founded.

As emerges very clearly from St Petersburg, the work carried out by the Bureau made the conferences appear time-consuming and even superfluous, given that the topics had already been faced in letters thanks to the surveys that had already taken place. Whatever, the fact that the Bureau had acquired a central role in the workings of the Union meant that at the same time Switzerland had consolidated its great power of influence over the Bureau. And the Bureau was to all effects and purposes a Swiss body: 1) its headquarters were in the Swiss capital; 2) it answered directly administratively and economically to the Federal Council; 3) the personnel were all Swiss nationals; 4) the various secretary generals were chosen from the heads of Swiss telegraphs. These associations, which had been developed during the earlier conference, were given definitive form at St Petersburg. The Federal Council, which had already come to the conference with much influence, came away from it with the knowledge that the power would remain in its hands for many years. In preserving the *status quo* resulting from the previous conferences, St Petersburg also confirmed the desire to entrust to Switzerland the role of guide and coordinator to the Union. Curchod played

et que, en fait de tarif, il n'y a eu aucun changement qui puisse porte atteinte aux intérêts du public suisse». *Message du Conseil fédéral à la haute Assemblée fédérale concernant la Convention télégraphique international de St. Petersburg*, pp. 13–14.

a decisive role as both a highly esteemed international telegraph expert who had been entrusted with the Bureau right from the beginning and an able Swiss manager who calmly and quietly, working mostly behind the scenes with a ceaseless diplomatic action creating compromises between opposing parties, had known how to bring the Bureau into Switzerland and then in later stages build up its power.

Conclusion

In the second half of the nineteenth century, Switzerland wielded great influence over the birth and development of the nascent international communications in four key moments.

Firstly between the mid-fifties and mid-sixties it was more active than any other nation in bringing closer the two "currents" of countries which had formed two distinct bodies for the running of international telegraphy. The Swiss were indeed the first to think of a common European space and then mediate between the member countries of the Austro-German Telegraph Union (UTAT) and the Western European Telegraph Union (WETO). Switzerland worked steadily and patiently at bringing the conventions of the two unions closer together until they overlapped.

A second step, which can be considered the natural outcome of this process, was the creation of the International Telegraph Union in 1865, today known as the International Telecommunication Union. It was the final realization of an idea and a strategy masterminded 10 years earlier by two heads of Swiss Telegraphs, who became such key players in the new organization: Karl F. Brunner-von Wattenwyl, who then moved over to Austrian Telegraphs, and Louis Curchod.

A third fundamental moment in which Switzerland guided the organization of international telecommunications was when the following 1868 Vienna Conference in Vienna approved Louis Curchod's plan to create a permanent body to supervise the international network. It took the name of *Bureau International des Administrations Télégraphique*, while Curchod became its first head. The headquarters were established in Bern and the administration entrusted to the Swiss Posts Department, which was the equivalent of putting it under the national government. The Bureau was to all effects and purposes a Swiss-controlled body up to WWII. That because: 1) its headquarters were kept in the Swiss capital and, later, in Geneva; 2) it was under the direct administrative and economic control of the Swiss government; 3) the personnel were all Swiss nationals; 4) the heads (secretaries general) were chosen from the top managers in the Swiss Posts.

After defending itself from attempts to subtract the new institution from Swiss influence, Switzerland brought the process to its end in the 1875 St Petersburg International Conference. The Bureau showed its clout in the way it prepared and had approved a new and definitive international convention which regulated world telecommunications policies until 1932, when the Union (together with telephones) underwent a radical transformation, indicated in the new name of International Telecommunication Union.

Right through the period Switzerland carried out a role of guide and control over international communications via means of the Bureau. It reached this position of power with a naturalness and gradualism which cannot fail to surprise those who read contemporary documents. In reality, and this is the principle point to make in this conclusion, a small state in the middle of Europe like Switzerland managed to reach this difficult goal of an international character thanks to political economic and technical features which the country managed to exploit and export to the rest of the continent. Neutrality, pacifism, the capacity for international mediation, "governmental internationalism" in foreign politics, the federal structure of the government, a central position in Europe, economic liberalism, technological skills. All these specialisms were determining in gaining control of the Bureau.

First, Switzerland was awarded permanent neutrality back at the Congress of Vienna in 1815, which was all-important for international telegraphy in at least two moments. The first was in the late fifties, while most nations were in conflict in their rush after land grabbing, Switzerland managed to mediate with politically isolated countries like Austria. Second and even more significantly, neutrality and pacifism were key qualities in the eyes of the Vienna Conference, which decided to assign the Bureau to Switzerland because of its "political position" and the Bureau being headquartered in a "neutral city" like Bern. Side by side with neutrality was the quality of pacifism, also innate in the national spirit for centuries, which had no doubt contributed to the creation of the International Red Cross Committee. The principle was indeed recalled in the First Paris Conference as one of the goals of the Telegraph Union.

A second element of the political nature, closely associated with the first and which played a determining role, was the capacity of the good and great to carry out missions of international mediation. One of the first cases of this natural tendency was in the months running up to the Paris conference when France did not invite Austria and the birth of the Union was in doubt. Switzerland acted both diplomatically and via its delegates (Curchod assured Brunner his support) and so managed without any conspicuous action to get Austria to the Conference. Here again, Switzerland showed it was suitable for the role of inter-European mediation because of another of its distinctive features: its multi-lingualism. Mediating meant constantly bringing together a French-speaking and a German speaking universe. Another example of international mediation emerged in the months following Curchod's resignation from the Bureau. Although the Federal Council could have appointed a successor, it decided to await the opinion of the Rome Conference and appointed Lendi the head of Swiss Telegraphs *interim*. This formally perfectly correct action also signalled the road to take and would be decisive in the moment in which Germany attacked the legitimacy of Swiss power over the Bureau. The Swiss ability in mediating and moving cautiously both favoured the birth of the TU and contributed to the conservation and final consecration of Swiss control over the Bureau.

A third strand of the Swiss policy is linked directly with the birth and control of the International Union. "Governmental internationalism" was a foreign policy strategy which Switzerland activated in this period in order to enhance its own relevance and power, and with this the running of international organizations was important. The TU was, as is well known, only the first to be placed on Swiss soil and put under the control of the government. It was the first practical application of a national policy whose extremely positive results acted as a trail-blazer for numerous other similar experiences. Headquartered in Switzerland (particularly Geneva) besides the International Red Cross have been or are the Universal Postal Union (UPU), the Bureau International de Travail (BIT), the Society of Nations (later the United Nations), the International Olympic Committee (CIO, now in Lausanne), the Organization of the United Nations (ONU), the United Nations Food and Agriculture Organization (FAO) the Federation International de Football Association (FIFA) and the World Trade Organization (WTO).

A fourth feature, that of the political system, was taken directly as a model for the design of the international office. In presenting his project to the Federal Council Curchod indicated he had drawn inspiration from the *Gouvernement du Vorort,* the Swiss government model preceding national unity. Some of its bodies like conferences or commissions had itinerant seats for essentially political reasons, as was the case with the government itself, which every two years moved from Canton to Canton. However, as the Bureau dealt with ordinary administration, it needed a permanent seat as had happened with the Federal Chancellery. It too needed archive space and had to be able to count on a stable bureaucratic apparatus. Federalism, one of the great strengths of the Swiss political model, became a template for the running of the European telegraph network, and most generally exported the Swiss model of unity in difference on a wider European scale.

Another feature in favour of the Swiss control of the Union was its particular geographical position. Being at the centre if Europe meant being automatically at the centre of the international telegraph network, providing Austria too was a member. Its geographical position was indeed one of the motives which convinced the Vienna Conference that the Swiss Administration would be the most suitable for running the service in the general interest, almost as though an equidistance between the various countries could correspond to a kind of administrative equidistance.

Another national specialism that was exported on to a European level was the liberal economic model promoted by the Radicals who had reached power post 1848. Based on the free circulation of goods, this model was applied in at least three issues concerning international telegraphy. Firstly, especially during the Paris Conference, Switzerland proposed a general reduction in telegraph charges so as to encourage use and make it accessible to a wider range of citizens. Secondly, in the debate over anti-dumping presented at Vienna, Curchod managed to get approval for an amendment which reconciled the freedom of single states to lower tariffs with a barrier blocking an indiscriminate cuts which would bring down the quality of the service. Then the liberal policy made the Swiss delegates side against the introduction of urgent telegrams deemed illiberal forms of communication which would be against universal access to all citizens. These three applications of the basic principle of Swiss Radical policy concealed a series of economic benefits for Switzerland. Cheaper tariffs would stimulate transit across the country and therefore increase flows into the country's coffers. The introduction of an anti-dumping would allow Switzerland to reach the position of privilege and block possible competition, as happened in 1865. Urgent telegrams would have raised cost, forced the government to take on extra staff and provide additional structures to deal with the increase in work. The constant worry about gaining advantages from the international rules in the making and have them pass for the general good is another under surface feature in the instructions given to the delegates at the international conferences. Many of the interventions of the Swiss delegates analysed in the present study can be read in this way, and prove the consummate level of diplomatic skill.

A last but still basic Swiss specialism was decisive in acquiring a position of control over international telegraphs: the recognition that Switzerland had a kind of technological superiority. Not by chance, the Vienna Conference decided to give the Bureau to Switzerland partly "to pay homage to the precision of the telegraph service of which Switzerland offers such a good example". However, the most important example of the weight given to competence in telegraphy is to be found in the figure of Curchod. The head of telegraphs was part of European technical elite, he exerted most influence in the decisions taken in the international conferences, and was almost unanimously elected head of his own creation, the Bureau, recognised as "a man joining experience in telegraph services and business with a recognised merit and even knowing several languages". Even the

Vienna Conference's choice to locate the Bureau in Switzerland was due to the esteem Curchod enjoyed and so this can be considered a decisive factor that put the control and management of the international telegraph network in Swiss hands.

In the ten years preceding and following the creation of the TU, Switzerland exploited and at times exported abroad some of its political, economic and technical specialisms which allowed it to gain control over nascent body. The TU was also a place which Switzerland experimented for the first time its own capacity to wield influence on an international level, run a supranational body and apply some of its famous political, economic and technical mainstays on a continental level.

References

Altermatt, C. "On Special Mission: Switzerland and its Diplomatic System." In *The Diplomats' World. A Cultural History of Diplomacy, 1815–1914*, edited by M. Mösslang and T. Riotte, 317–344. Oxford: Oxford University Press, 2008.

Anceau, É. *Napoléon III. Un Saint-Simon à cheval*. Paris: Tallandier, 2008.

Antonetti, G. *Louis-Philippe*. Paris: Librairie Arthème Fayard, 1994.

Badenoch, A., and A. Fickers. "Introduction: Europe materializing? Toward a transnational history of European infrastructures." In *Materializing Europe: transnational infrastructures and the project of Europe*, edited by A. Badenoch and A. Fickers, 1–23. Basingstoke, Hampshire and New York: Palgrave Macmillan, 2010.

Balbi, G., Fari, S., Calvo S., and G. Richeri. "Swiss specialties: Switzerland's role in the genesis of the Telegraph Union, 1855–1875." *Journal of European Integration History* 19/2 (2013): 207–225.

Balbi, G., Fari, S., Calvo S., and G. Richeri. "Specialità svizzere. L'influenza della Confederazione elvetica sull'origini dell'Unione Telegrafica, 1855–1875." *Tst: Transportes, Servicios y telecomunicaciones* 25 (2013): 150–175.

Balbi, G., Calvo S., Fari, S., and G. Richeri. "'Bringing together the two large electric currents that divide Europe': Switzerland's Role in Promoting the Creation of a Common European Telegraph Space, 1849–1865." *ICON* 15 (2009): 61–80.

Balbi, G. "Studying the Social History of Telecommunications. Between Anglophone and Continental Tradition." *Media History* 15/1 (2009): 85–101.

Balbi, G. *Le origini del telefono in Italia. Politica, economia, tecnologia e società*. Milan: Bruno Mondadori, 2011.

Balbi, G. "Telecommunications." In *Handbook of Communication History*, edited by P. Simonson, J. Peck, R. T. Craig, and J. P. Jackson, 209–222. London and New York: Routledge, 2012.
Balbi, G. "The Origins of the Telephone in Italy, 1877–1915: Politics, Economics, Technology and Society." *International Journal of Communication* 5 (2011): 1058–1081.
Barbier, F. et C. Bertho-Lavenir. *Histoire des medias: de Diderot à Internet*. Paris: A. Colin, 2003.
Beauchamp, K. *History of Telegraphy*. IEE: Stevenage, 2001.
Bergier, J-F. *Wirtschaftgeschichte der Schweiz*. Zürich: Benziger Verlag, 1983.
Bertho-Lavenir, C. (dir.). *Histoire des télécommunications en France*. Toulouse: Eres, 1984.
Bignell, J., and A. Fickers (eds.). *A European television history*. Malden MA and Oxford: Wiley-Blackwell, 2008.
Blackbourn, D. *The Long Nineteenth century: a History of Germany, 1780–1918*. New York: Oxford University Press, 1998.
Blatherwick, D. E. S. *The international politics of telecommunications*. Berkeley, CA: Institute of International Studies, University of California, 1987.
Bourdon, J. *Du service public à la télé-réalité. Une histoire culturelle des télévisions européennes, 1950–2010*. Paris: INA Éd., 2011.
Brauns-Packenius, O. "Die Internationale Telekommunikations-Union im Wandel der Zeit." *Archiv für deutsche Postgeschichte* 1 (2001): 46–54.
Breguet, E. *Breguet horloger depuis 1775, vie et postérité d'Abraham-Louis Breguet*. Paris: ed. Alain de Gourcuff, 1997.
Brooks, J. *Telephone. The First Hundred Years*. New York & London: Harper & Row, 1972.
Burns, R. *Communications: an international history of the formative years*. London: IEE, 2004.
Calvo, S., Balbi, G., Fari, S., and G. Richeri. "La voie suisse aux télécommunications. Politique, économie, technologie et société (1850–1915)." *Revue Suisse d'Histoire* 61/4 (2011): 435–453.
Carrè, P. A. "Archeologie d'une Europe des télécommunications." *Revue Française des Télécommunications* 70 (1989): 72–83.

Chamberlain, M. E. *The Scramble for Africa*. London: Longman, 1974.
Cipolla, C. M. *Le macchine del tempo: l'orologio e la società, 1300–1700*. Bologna: il Mulino, 1980.
Codding, G. A. *The International Telecommunication Union. An Experiment in international cooperation*. Leiden: E. J. Brill, 1952.
Codding, G. A. *The Universal Postal Union: coordinator of the international mails*. New York: New York University Press, 1964.
Codding, G. A., and A. M. Rutkowski. *The International Telecommunication Union in a Changing World*. Deedham: Artech House, 1982.
Conrad, C. "Social policy history after the transnational turn." In *Beyond Welfare State Models. Transnational Historical Perspectives on Social Policy*, edited by P. Kettunen and K. Petersen, 218–240. Cheltenham, UK and Northampton, MA: Edward Elgar, 2011.
Cornwall M. (ed.) *The last years of Austria-Hungary: a multi-national experiment in early twentieth-century Europe*. Exeter: University of Exeter Press, 1990.
Cowhey, P. F. "The International Telecommunications Regime: The Political Roots of Regimes for High Technology." *International Organization* 44/2 (1990): 169–199.
De Wit, O. *Telefonie in Nederland 1877–1940*. Den Haag: Cramwinckel, 1998.
Descalzi, A. "Creacion y desarollo de la Union Internacional de telecomunicaciones." In *Las comunicaciones entre Europa y América, 1500–1993*, edited by B. Magro, M. Llorente, O. Carvajal, 287–298. Palacio de Congresos de Madrid: Actas del I Congreso Internacional de Comunicaciones, 30 de noviembre – 3 de diciembre.
Deutsch, W. *Habsburgs Rückzug aus Italien: die Verhandlungen von Villafranca und Zurich 1859*. Wien und Leipzig: Adolf Luser, 1940.
Devin, G., and M.-C. Smouts. *Les organisations internationales*. Paris: Colin, 2011.
Direction Générale des PTT (publ. par). *Un siècle de télécommunications en Suisse, 1852–1952. Tome I, Télégraphe*. Berne, 1952.
Direction Générale des PTT (publ. par). *Un siècle de télécommunications en Suisse, 1852–1952. Tome II, Téléphone, sources de courant et installations d'énergie, lignes aériennes et lignes souterraines*. Berne, 1959.

Direction Générale des PTT (publ. par). *Un siècle de télécommunications en Suisse, 1852–1952. Tome III, Téléphone (suite et fin).* Berne, 1963.

Dordick, H. S. "The origins of the universal services: History as a Determinant of Telecommunications Policy." *Telecommunications Policy* 14 (1990): 223–231.

Droz, J. (dir.) *Histoire de l'Allemagne. Vol 1. La Formation de l'unité allemande, 1789–1871.* Paris: Hatier, 1970.

du Bois, P. *Le Guerre du Sonderbund. La Suisse de 1847.* Paris: Éditions Alvik, 2002.

Durand Barthez, P. *Union Internationale des Télécommunications.* Thèse pour le doctorat en droit, Université de Paris I-Pantheon-Sorbonne Sciences Economiques-Science Humaines-Sciences Juridiques, 1979.

Ehs, T. "An Unwritten History: The Europeanisation of Switzerland." Paper presented at the EUSA 10th Biennial Conference, Montreal, Canada, May 17–19, 2007.

Ehs, T. "History in Europeanisation Studies: Lessons from Switzerland." *Contemporary European Studies* 3/1 (2008): 23–39.

Extramiana, J. *Historia de las guerras carlistas.* San Sebastián: L. Haranburu, 1979–1980.

Fari, S. *Una penisola in comunicazione. Il servizio telegrafico italiano dall'Unità alla Grande Guerra*. Bari: Cacucci Editore, 2008.

Fari, S. "Technology on the wire. Technological changes in the first thirty years of the Italian telegraph experience: achievements and difficulties." In *Communication and its lines. Telegraphy in the 19th Century among economy, politics and technology*, edited by A. Giuntini, 135–158. Prato: Istituto di Studi Storici Postali, 2004.

Fari, S., Balbi, G., and G. Richeri. "Telecommunications Italian Style. The shaping of the constitutive choices (1850–1914)." *History of Technology*, forthcoming 2014.

Fari, S. "Uccide più la parola che la spada. Telecomunicazioni e questioni militari nell'Italia del XIX secolo." *Ricerche Storiche* 1 (2006): 5–28.

Fari, S., Balbi, G., and G. Richeri. "Last in First Out. How the Telegraph Brought to International Organization Before Post and Trains." Paper presented at the ICOHTEC Conference, Barcelona, July 10–14, 2012.

Fari, S., Balbi, G., and G. Richeri. "'European Multilateralism' (1848–1865): a Telegraphic idea?" Paper presented at the 6th Plenary Conference of Tensions of Europe, Paris, September 19–21, 2013.

Fari S., Balbi G., and G. Richeri. "A common technical culture of telegraphy: the Telegraph Union and the significance of Technological Standardization, 1865–1875." Paper presented at the HISTELCON Conference, Pavia, September 5–7, 2012 (available <http://ieeexplore.ieee.org/xpl/articleDetails.jsp?tp=&arnumber=6487580&queryText%3DHistelcon>).

Feldman, M. L. B. *The United States in the International Telecommunication Union and in Pre-ITU conferences: Submarine Cables, Overland Telegraph, Sea and Land Radio, Telecommunications.* Baton Rouge: printed autonomously, 1974.

Fickers, A., and P. Griset. *Eventing Europe: Electronic Information and Communication Spaces in Europe, 1850–2000*. London: Palgrave/MacMillan, forthcoming 2014.

Fickers, A., and S. Lommers. "Eventing Europe: broadcasting and the mediated performances of Europe." In *Materializing Europe: transnational infrastructures and the project of Europe*, edited by A. Badenoch and A. Fickers, 225–251. Basingstoke, Hampshire and New York: Palgrave Macmillan, 2010.

Fischer, C. S. *America Calling: A Social History of the Telephone to 1940*. Berkeley: University of California press, 1992.

Fischer, A. "Swiss telecommunications policy: From state monopoly to intense regulation." *Flux* 2-3/72 (2008): 78–91.

Flichy, P. *Une histoire de la communication moderne: espace public et vie privée*. Paris: La Decouverte, 1991.

Foner, E., and O. Mahoney. *America's Reconstruction: people and politics after the Civil War*. New York: HarperCollins, 1995.

Foreman-Peck, J. "L'état et le développement du réseau de télécommunications en Europe à ses débuts." *Histoire, économie et société* 3 (1989): 383–402.

Gilliard, C. *Histoire de la Suisse*. Paris: PUF, 1987.
Giuntini, A. *Le meraviglie del mondo: il sistema internazionale delle comunicazioni nell'Ottocento*. Prato: Istituto di studi storici postali, 2011.
Griset, P. *Technologie, entreprise et souveraineté: les télécommunications transatlantiques de la France, 1869–1954*. Lille: A.N.R.T. Université de Lille III, 1993.
Grossi, V. "Le rôle international de personnalités suisses du XIXe siècle dans le domaine des télégraphes." *Hispo* (1984): 43–50.
Grossi, V. "Technologie et diplomatie suisse au XIXe Siècle." *Relations internationales* 39 (1984): 287–307.
Gugerli, D., Kupper, P., und D. Speich. *Die Zukunftsmaschine. Konjunkturen der ETH Zürich 1855–2005*. Zürich: Chronos Verlag, 2005.
Habermas, J. *The structural transformation of the public sphere: an inquiry into a category of bourgeois society*. Cambridge MA: MIT Press, 1991.
Hamelink, C. J. *The politics of world communication: a human rights perspective*. London and Thousand Oaks, CA: Sage, 1994.
Hampf, M., and S. Müller-Pohl (eds.) *Global Communication Electric. Business, News and Politics in the World of Telegraphy*. Frankfurt and New York: Campus, 2013.
Hazlewood, A. "The Origin of the State Telephone Service in Britain." *Oxford Economic Papers, new series* 5/1 (1953): 13–25.
Headrick, D. *When Information Came of Age. Technologies of Knowledge in the Age of Reason and Revolution, 1700–1850*. London & New York: Oxford University Press, 2000.
Headrick, D. R. *The invisible weapon. Telecommunications and international politics 1851–1945* New York: Oxford University Press, 1991.
Helgesson, C.-F. *Making a natural monopoly: the configuration of a techno-economic order in Swedish telecommunications*. Stockholm: Stockholm School of Economics, the Economic Research Institute, 1999.
Herren, M. *Internationale Organisationen seit 1865. Eine Globalgeschichte der internationalen Ordnung*. Darmstadt: Wissenschaftliche Buchgesellschaft, 2009.

Herren, M. "Governmental Internationalism and the Beginning of a New World Order in the Late XIX Century." In *The Mechanics of Internationalism: Culture, Society, and Politics from the 1840s to the First World War*, edited by M. H. Geyer and J. Paulmann, 121–144. Oxford: Oxford University Press, 2001.

Herzig, A. *Schlesien. Das Land und seine Geschichte in Bildern, Texten und Dokumenten*. Hamburg: Ellert & Richter Verlag, 2008.

Hills, J. *The struggle for control of global communication: the formative century*. Urbana: University of Illinois Press, 2002.

Hofmann, M. "The Telephone and Its Significance to Switzerland's First Subscribers." *Bulletin Technique PTT* 10 (1980): 390–392.

Horrenberger, J. *L'Union Internationale des Télécommunications ou Les exigences techniques comme factor de la coopération internationale*. Mémoire pour l'obtention du diplôme des Hautes Etudes Européens section des sciences de l'information, Université de Strasbourg, 1976.

Hugill, P. J. *Global communications since 1844: geopolitics and technology*. Baltimore: Johns Hopkins University Press, 1999.

Huurdeman, A. A. *The Worldwide History of Telecommunications*. London: IEE, 2003.

Ibsen, K. *Maximilian, Mexico, and the Invention of Empire*. Nashville: Vanderbilt University Press, 2010.

Imlah, A. G. *Britain and Switzerland 1845–60: a study of Anglo-Swiss relations during some critical years for Swiss neutrality*. Hamden, Connecticut: Archon Books, 1966.

Iriye, A. P., and Y. Saunier (eds.). *The Palgrave dictionary of transnational history*. Basingstoke UK and New York: Palgrave Macmillan, 2009.

ITU. *L'Union Internationale des télécommunications, cent ans de coopération international*. Genève, 1965.

ITU. *L'Union télégraphique international: 1865–1915*. Berne: Bureau international de l'Union télégraphique, 1915.

John, R. R. *Network Nation. Inventing American Telecommunications*. Cambridge MA and London: The Belknap Press of Harvard University Press, 2010.

John, R. R. "Private enterprise, public good? Communications deregulation as a national political issue, 1839–1851." In *Communication and its lines. Telegraphy in the 19th Century among economy, politics and technology*, edited by A. Giuntini, 33–56. Prato: Istituto di Studi Storici Postali, 2004.

John, R. R. "Telecommunications." *Enterprise and Society* 9 (2008): 507–520.

Jost, H. U. "Origines, interprétations et usages de la 'neutralité helvétique'." *Matériaux pour l'histoire de notre temps* 93 (2009): 5–12.

Kieve, J. *The Electric Telegraph. A Social and Economic History*. Devon: David and Charles Newton Abbot, 1973.

Kinnebrock, S., Schwarzenegger, C., and E. McLuskie (eds.). *What is Communication History? European Answers II*. Special Issue of *Medien & zeit* 3–4 (2011).

Kitchen, M. *A history of modern Germany: 1800 to the present*. Chichester, West Sussex and Malden, MA: Wiley-Blackwell, 2012.

Kobelt, C. "100 years of Telephone Service in Switzerland." *Bulletin Technique PTT* 10 (1980): 344–363.

Kott, S. (dir.). *Une autre approche de la globalisation: Socio-histoire des organisations internationales (1900–1940)*. Numéro spécial de *Critique internationale* 52 (2011).

Kreis, G. *Der Weg zur Gegenwart. Die Schweiz im neunzehnten Jahrhundert*. Basel: Basel Birkhäuser, 1986.

Kriesi, H. *Le système politique suisse*. Paris: Economica, 1998.

L'Union internationale des télécommunications, 1865–1995. London: International Systems and Communications LTD, 1995.

Laborie, L. *L'Europe mise en réseaux. La France et la coopération internationale dans les postes et les télécommunications, années 1850-années 1950*. Bruxelles: PIE Peter Lang, 2010.

Laborie, L. "De quoi l'universel est-il fait? L'Europe, les empires et les premières organisations internationales." *Les Cahiers Irice*, 9 (2012): 11–22.

Laborie, L. "En chair et en normes. Les participants aux conférences de l'Union internationale des télécommunications, de sa fondation à sa refondation (1865–1947)." *Flux. Cahiers scientifiques internationaux Réseaux et Territoires* 74 (2008): 92–98.

Laborie L. *La France, l'Europe et l'ordre International des communications, 1865–1959.* Thèse pour le Doctorat en histoire contemporaine, Université Paris IV-Sorbonne, 2006.

Lewis, W., Balderstone, S., and J. Bowan. *Events That Shaped Australia.* Chatswood, N.S.W.: New Holland, 2006.

Liniger-Goumaz, M. *La Suisse, sa neutralité et l'Europe.* Genève: Éditions du Temps, 1964.

Lommers, S. B. *Europe – on air: interwar projects for radio broadcasting.* Eindhoven: Technische Universiteit Eindhoven, 2012.

Lyall, F., and P. B. Larsen. *Space law: a treatise.* Farnham, Surrey, UK and Burlington, VT: Ashgate, 2009.

Lyall, F. *International communications: the International Telecommunication Union and the Universal Postal Union.* Farnham, Surrey UK and Burlington, VT: Ashgate, 2011.

Mack Smith, D. *Modern Italy: A Political History.* Ann Arbor: University of Michigan Press, 1997.

Mahajan, S. *British foreign policy, 1874–1914: the role of India.* London and New York: Routledge, 2002.

McCullough, D. *The Path between the Seas: The Creation of the Panama Canal, 1870–1914.* New York: Simon and Schuster, 1977.

Meyer, V. *L'Union Internationale des Telecommunications et son Bureau.* Bern: manuscript, 1946.

Michaelis, A. R. *Du sémaphore au satellite.* Genève: ITU, 1965.

Milivojević, M., and P. Maurer (eds.). *Swiss neutrality and security: armed forces, national defence, and foreign policy.* New York: Berg and St. Martin's Press, 1990.

Milza, P. *Napoléon III.* Paris: Perrin, 2006.

Misa, T. J., and J. Schot. "Inventing Europe: Technology and the Hidden Integration of Europe. Introduction to the special issue." *History and Technology* 21/1 (2005): 1–20.

Müller-Pohl S. "Working the Nation State: Submarine Cable Actors, Cable Transnationalism and the Governance of the Global Media System, 1858–1914." In *The Nation State and Beyond. Governing*

Globalization Processes in the Nineteenth and Twentieth Century, edited by I. Löhr and R. Wenzlhuemer. Berlin: Springer Verlag Berlin, 2013.

Müller-Pohl, S., and M. Hampf (eds.). *Global Communication Electric: Actors of a Globalizing World.* Frankfurt: Campus, 2013.

Nachszunow, G. *Développement des télécommunications et organisations internationales*. Créteil: W. Nachszunow, 1989.

Navarro, J. F. *ITU: síntesis de organización, funcionamiento y objetivos de la Union Internacional de Telecomunicaciones.* Chile: Entel-Chile, 1975.

Noam, E. *Telecommunications in Europe*. New York: Oxford University Press, 1992.

Perrin, G. *La neutralité permanente de la Suisse et les organisations internationales*. Heule: UGA, 1964.

Perry, C. R. *The Victorian Post Office. The growth of a bureaucracy*. Woodbridge: The Royal Historical Society/The Boydell Press, 1992.

Peters, A., and S. Peter. "International Organizations: Between Technocracy and Democracy." In *The Oxford handbook of the history of international law*, edited by B. Fassbender and A. Peters, 170–197. Oxford: Oxford University Press, 2012.

Pflieger, G. "La Suisse est-elle un espace première classe?" *Flux* 2-3/72 (2008): 4–12.

Pravato, S. "De Télécom PTT à Swisscom." In *Du monopole à la concurrence. Analyse critique de l'évolution de 6 entreprises suisses*, dir. M. Finger, S. Pravato, J.-N. Rey. Lausanne: LEP, 1994.

Quinn, A. *Hell with the Fire Out: A History of the Modoc War.* Boston: Faber & Faber, 1997.

Rapold, H. *Strategische Probleme der schweizerischen Landesverteidigung im 19. Jahrhundert*. Frauenfeld: Huber, 1951.

Reindl, J. *Der Deutsch-Österreichische Telegraphenverein und die Entwicklung des Deutschen Telegraphenwesens 1850–1871*. Frankfurt am Main: Peter Lang, 1993.

Reindl, J. "Partikularstaatliche Politik und Technische Dynamik: Die Drahtgebundene Telegraphie und der Deutsch-Österreichische

Telegraphenverein von 1850." In *Vom Flügeltelegraphen zum Internet: Geschichte der modernen Telekommunikation*, edited by H.-J. Teuteberg and C. Neutsch, 27–46. Stuttgart: F. Steiner, 1998.

Richeri, G. "The media amidst the enterprises, the public and the State." *Studies in Communication Sciences* 6/2 (2006): 131–143.

Roh, H. *Fédéralisme politique et décentralisation économique et industrielle*. Thèse pour le doctorat, Université de Paris, 1960.

Rosenberg, N. *Exploring the black box: technology, economics, and history*. Cambridge UK and New York: Cambridge University Press, 1994.

Roth, F. *La guerre de 1870*. Paris: Fayard, 1990.

Ruffieux, R. "La Suisse des Radicaux 1848–1914." In *Nouvelle Histoire de la Suisse et des Suisses*, dir. G. Andrey et al., vol. 3. Payot: Lausanne, 1982–1983.

Rutkowski, A., and W. P. Dizard. "The International Telecommunication Union and the United States: partners of rivals?" In *International Telecommunications and information Policy: Papers from a Symposium (P.1) and a report of the National Telecommunications Information Administration (NTIA) 1983 (P.2)*, 27–45. Washington: Communications Pr., 1984.

Sanguin, A.-L. *La Suisse, essai de géographie politique*. Gap: Editions Ophrys, 1983.

Schiffino, N. *Crises politiques et démocratie en Belgique*. Paris: L'Harmattan, 2003.

Schipper, F., and J. Schot. "Infrastructural Europeanism, or the project of building Europe on infrastructures." *History and Technology* 27/3 (2011): 245–264.

Schmidt, S., and R. Werle, *Coordinating Technology: Studies in the International Standardization of Telecommunications*. Cambridge-London: MIT Press, 1998.

Schneider, V. *Die Transformation der Telekommunikation. Vom Staatsmonopol zum globalen Markt, 1800–2000*. Frankfurt am Main: Campus, 2001.

St. John, I. *Gladstone and the logic of Victorian politics*. London & New York: Anthem Press, 2010.

Starr, P. *The creation of the media: political origins of modern communications*. New York: Basic Books, 2004.

Sterling, C., Bernt, P., and M. B. H. Weiss. *Shaping American telecommunications: a history of technology, policy and economics*. Mahwah, New York: Lawrence Erlbaum Associates, 2006.

Tanner, A. "Das Recht auf Revolution." In *Im Zeichen der Revolution: der Weg zum schweizerischen Bundesstaat, 1798–1848*, edited by T. Hildbrand and A. Tanner. Zürich: Chronos, 1997.

Tarjanne, P. (avant-propos par). *Union internationale des télécommunications: célèbre 130 années d'existence 1865–1995*. Genève: International Telecommunication Union and International Systems and Communications Ltd., 1996.

Taylor, A. J. P. *Bismarck: The Man and the Statesman*. London: Hamish Hamilton, 1955.

Tegge, A. *Die Internationale Telekommunikations-Union: Organisation und Funktion einer Weltorganisation im Wandel*. Baden-Baden: Nomos Verlagsgesellschaft, 1994.

Thompson, J. B. *The media and modernity: a social theory of the media*. Stanford, CA: Stanford University Press, 1995.

Thomson, G. *The birth of modern politics in Spain: democracy, association and revolution, 1854–75*. Basingstoke and New York: Palgrave Macmillan, 2010.

Van der Vleuten, E., and A. Kaiser (eds.). *Networking Europe. Transnational Infrastructures and the Shaping of Europe, 1850–2000*. Sagamore Beach, MA: Science History Publications, 2006.

Van der Vleuten, E. "Towards a transnational history of technology: meanings, promises, pitfalls." *Technology and Culture* 49/4 (2008): 974–994.

Wawro, G. *The Austro-Prussian War: Austria's war with Prussia and Italy in 1866*. Cambridge and New York: Cambridge University Press, 1996.

Weber, M. *Economy and society; an outline of interpretive sociology*. New York: Bedminster Press, 1968.

Weibel, E. "La coexistence linguistique face au défi européen." *Annales de la Faculté de Droit et de Science Politique de l'université de Clermont* 25 (1989).

Wenzlhuemer, R. *Connecting the Nineteenth-Century World. The Telegraph and Globalization*. Cambridge UK: Cambridge University Press, 2013.
Wenzlhuemer, R. "Editorial – Telecommunication and Globalization in the Nineteenth Century." *Historical Social Research / Historische Sozialforschung* 35/1 (2010): 7–18.
Winsek, D. R., and R. M. Pike. *Communication and Empire: Media, Markets, and Globalization, 1860–1930*. Durham: Duke University Press, 2007.

Kolibri 360 GmbH
Ferdinand-Jühlke-Straße 7
99095 Erfurt, Deutschland
produktsicherheit@kolibri360.de